Micro-Joining of Materials

This book addresses primary and secondary procedures for combining sophisticated materials such as metals and alloys, intermetallic, ceramics, glasses, polymers, superalloys, electronic materials, and composites in similar and dissimilar combinations, based on the most recent research. It further explores the current and future micro-joining techniques that are rapidly expanding, including delicately and precisely combining microscale components and materials. It offers a detailed reference to the status of micro-joining and potential future possibilities.

- Provides an understanding of micro-joining techniques with a focus on its various applications such as in biomedical, electrical and electronics, and aerospace
- Covers specialized techniques such as micro-electronics wire bonding, solid-state diffusion bonding, diffusion soldering, laser soldering, and laser micro-welding
- Details analytical and numerical modeling of micro-welding processes with their optimization
- Describes complete mechanism of joining of various parts/components
- Includes case studies with present day daily life applications

This book is aimed at researchers and graduate students in materials science and microfabrication.

Advanced Materials Processing and Manufacturing

Series Editor: Kapil Gupta

The CRC Press Series in Advanced Materials Processing and Manufacturing covers the complete spectrum of materials and manufacturing technology, including fundamental principles, theoretical background, and advancements. Considering the accelerated importance of advances for producing quality products for a wide range of applications, the titles in this series reflect the state-of-the-art in understanding and engineering the materials processing and manufacturing operations. Technological advancements for enhancement of product quality, process productivity, and sustainability, are on special focus including processing for all materials and novel processes. This series aims to foster knowledge enrichment on conventional and modern machining processes. Micro-manufacturing technologies such as micro-machining, micro-forming, and micro-joining, and Hybrid manufacturing, additive manufacturing, near net shape manufacturing, and ultra-precision finishing techniques are also covered.

Fused Deposition Modelling: Technology, Applications, and Developments
Edited by Anshuman Kumar Sahu and Siba Sankar Mahapatra

Manufacturing Tribology
Edited by Jibin T Philip, Chacko Preno Koshy, and MD Mathew

Machine Learning for Advanced Manufacturing
Edited by Nishant Ranjan, Rashi Tyagi, Ranvijay Kumar, Ashutosh Tripathi, and Amit Verma

Engineered Ceramics and Composite Materials: From Fundamentals to Applications
Edited by Neethu Bhaskar, Sathish Kumar Palaniappan, Sanjay Mavinkere Rangappa, and Suchart Siengchin

Additive Manufacturing of Thin-Walled Structures
Edited by Sagar Nikam, Mayur Sawant and Shaun McFadden

Developments in Electric Discharge Machining
Edited by Ranjan Kumar Ghadai, Prosun Mandal and Sujeet Kumar Chaubey

Micro-Joining of Materials: Current and Future Needs
N. Rajesh Jesudoss Hynes, P. Shenbaga Velu, N. J. Vignesh, R. Sankaranarayanan, and J. Angela Jennifa Sujana

For more information about this series, please visit: www.routledge.com/Advanced-Materials-Processing-and-Manufacturing/book-series/CRCAMPM

Micro-Joining of Materials

Current and Future Needs

N. Rajesh Jesudoss Hynes
P. Shenbaga Velu
N. J. Vignesh
R. Sankaranarayanan
J. Angela Jennifa Sujana

CRC Press
Taylor & Francis Group
Boca Raton London New York

CRC Press is an imprint of the
Taylor & Francis Group, an **informa** business

Designed cover image: Shutterstock

First edition published 2027
by CRC Press
2385 NW Executive Center Drive, Suite 320, Boca Raton FL 33431

and by CRC Press
4 Park Square, Milton Park, Abingdon, Oxon, OX14 4RN

CRC Press is an imprint of Taylor & Francis Group, LLC

ISBN: 978-1-032-79692-5 (hbk)
ISBN: 978-1-032-81989-1 (pbk)
ISBN: 978-1-003-50239-5 (ebk)

DOI: 10.1201/9781003502395

Typeset in Times
by SPi Technologies India Pvt Ltd (Straive)

Contents

About the Authors

N. Rajesh Jesudoss Hynes is a dedicated researcher in Opole University of Technology, Opole, Poland and an accomplished Professor and dedicated researcher in the Department of Mechanical Engineering of Mepco Schlenk Engineering College, Sivakasi, India, with over two decades of experience. Recognized among the Top 2% of Scientists worldwide by Stanford University and reputed publishing houses for six consecutive years, his expertise lies in sustainable material development, metal matrix composites, and advanced friction-based manufacturing processes. He is the recipient of the NAWA–ULAM fellowship of the Polish Government and Fast Track Young Scientist Award of Department of Science and Technology, India. To his credit, he has published 146 articles in international journals of repute and holds 20 patents for his innovation in the field of manufacturing. He has undertaken several research projects from national funding agencies such as Narodowa Agencja Wymiany Akademickiej, Warsaw; Department of Science and Technology, New Delhi; Defence Research Development Laboratory, Hyderabad; Micro Small and Medium Enterprises Department, New Delhi; Institution of Engineers, Kolkata; All India Council for Technical Education, New Delhi; Rural Technology Action Group (RuTAG), Indian Institute of Technology – Madras; and Tamil Nadu State Council for Science and Technology, Chennai.

P. Shenbaga Velu is currently serving as an Assistant Professor (Senior Grade) in the Centre for Advanced Materials and Innovative Technologies/School of Mechanical Engineering at Vellore Institute of Technology (VIT), Chennai, India. He holds a PhD in mechanical engineering from Anna University, Chennai, and has postdoctoral research experience at VIT Chennai. His research interests include advanced manufacturing, composite materials, joining of dissimilar materials, and additive manufacturing. Velu has authored over 60 international journal publications, 12 book chapters, and holds 6 patents. He is also an active reviewer for several SCI-indexed journals and a recognized PhD supervisor at VIT University.

N. J. Vignesh, is currently serving as an Assistant Professor in the Department of Mechanical Engineering at SRM Madurai College for Engineering and Technology, Pottapalayam, Sivaganga District, Tamil Nadu, India. He earned his doctorate in mechanical engineering from Anna University, Chennai, through Mepco Schlenk Engineering College (Autonomous), Sivakasi, where he also completed his master's degree in CAD/CAM with distinction. With over ten years of academic and research experience, Vignesh specializes in joining technologies, composite materials, and fiber metal laminates. He has authored and co-authored several research papers in reputed international journals and book chapters published by several reputed publishing houses. His innovative work has led to the granting of multiple patents in the field of mechanical design and manufacturing. A recognized PhD supervisor at Anna University, he actively contributes to academic development through teaching, research mentoring, and professional training programs. His current research focuses on sustainable materials and advanced joining processes for aerospace and automotive applications.

R. Sankaranarayanan is Assistant Professor in the Department of Mechanical Engineering, Sri Venkateshwara College of Engineering, Chennai. He possesses seven years of industrial experience and six years of teaching experience. He did postgraduation in CAD/CAM and a PhD in mechanical engineering (Welding of MMC) in Mepco Schlenk Engineering College, Sivakasi under Anna University, Chennai. His areas of interest include welding of metal matrix composites, joining of dissimilar materials and sustainable manufacturing. He has published more than 30 journal publications and 7 book chapters in his areas of interest.

J. Angela Jennifa Sujana is currently serving as Professor and Head in the Department of Artificial Intelligence and Data Science, Mepco Schlenk Engineering College, Sivakasi, India. She completed her PhD in the area of scientific workflow scheduling in cloud computing from M.S. University, Tirunelveli, India. Her area of research includes artificial intelligence, machine learning, cloud computing, soft computing, and fog and edge computing. She is a recognized supervisor at Anna University under Faculty of Information and Communication. She is the Principal Investigator for the projects titled "Energy Aware Smart Street Lamp Design for Autonomous maintenance using AI and Fog Computing" sponsored by DST, New Delhi and "Computer Guided Drill-less Dentistry Using Micro-Air Abrasion Technology" sponsored by TNSCST, Chennai. She is a recognized reviewer in several reputed journals. She is a certified EMC Academic Associate on "cloud infrastructure and services" and a certified deep learning expert by NVIDIA Academy.

Preface

The field of manufacturing and engineering is constantly evolving to satisfy the demands of contemporary industries in an era of rapid technological improvement. The advent of micro-joining techniques, which has recently attained notable prominence, is one of the crucial marks of this progression. The science and technology of micro-joining have become a support of modernization of manufacturing as the desire for fabrication of smaller, trickier, and more effective miniaturized components has increased around the world.

This book, *Micro-Joining of Materials – Current and Future Needs*, delves deeply into the dynamic world of micro-joining, examining its current status and snooping into the fascinating possibilities that lie ahead. The role of artificial intelligence in the domain of micro-joining is discussed, along with its barriers to implementation. This comprehensive collection distills the wisdom of contemporary researchers who solved the puzzles of small-scale assembly and intends to offer a broad overview of the approaches, difficulties, and prospective solutions in this developing field.

Micro-joining is a multidisciplinary field that draws influence from a variety of other fields, including biotechnology, medical devices, materials science, mechanical engineering, and electronics. The intrinsic intricacy of operating at the microscale provides a variety of difficulties, from guaranteeing accuracy and dependability to coping with unusual material characteristics. Researchers, engineers, and hobbyists are all motivated to investigate new approaches, materials, and technologies that could redefine the capabilities of micro-joining.

This book is organized to give the readers a walk through over the fundamentals of micro-joining, including developing approaches that challenge conventional wisdom, such as welding, soldering, adhesive bonding, and other methods. Each part of the book offers insights into the fundamental mechanics, demonstrates practical applications, and digs deeper into the intricacies that make micro-joining unique from its macroscopic equivalents. The book has a futuristic outlook, anticipating the opportunities presented by developments in nanotechnology, additive manufacturing, and smart materials.

It is crucial to recognize that this book is not a time capsule; rather, it depicts a certain moment throughout a trip that is constantly changing. Discoveries and innovations are sure to surface as the science of micro-joining continues to develop, altering our understanding and broadening the possibilities. Explore the complexity, problems, and possibilities associated with micro-joining processes in the pages that follow and be amazed by the marvels of recently developed techniques.

1 Introduction to Micro-Joining

Advanced joining technologies that can function at microscopic scales are becoming more and more necessary as a result of the miniaturization of components in contemporary industries like precision engineering, aerospace, biomedical devices, and microelectronics. The accuracy, thermal control, and material compatibility needed at the micro and nano levels are frequently not met by traditional joining techniques, despite their long history in macro-scale manufacturing. The creation of dependable, high-performance joints in components with dimensions usually less than 1 millimeter has led to the emergence of a specialized field known as micro-joining.

Micro-joining processes are sophisticated manufacturing methods used in precision engineering, biomedical devices, and microelectronics to assemble small parts.[1] Micro-joining works at the micrometer to millimeter level with very little heat input, in contrast to conventional joining, which works on a millimeter to meter scale and demands a high heat input (Figure 1.1). This reduces or eliminates material distortion while enabling extraordinarily high accuracy and precision, frequently within micrometers. As a result, micro-joining enables reliable bonding of delicate and heat-sensitive materials, making it a critical technology for modern micro-scale fabrication.

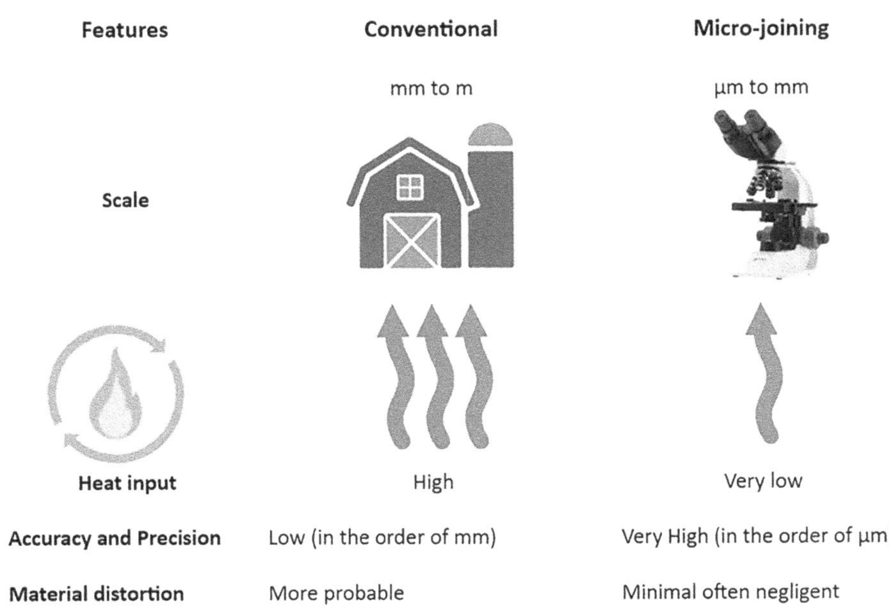

FIGURE 1.1 An overview of micro-joining processes compared to conventional processes.

DOI: 10.1201/9781003502395-1

1.1 EVOLUTION OF MICRO-JOINING PROCESSES

The invention of precision soldering and welding methods in the electronics sector in the middle of the 20th century marked the arrival of micro-joining techniques (Figure 1.2). At first, techniques like solder reflow and resistance spot welding were modified for micro-scale assemblies. But the growing complexity of materials and geometries led to the creation of increasingly sophisticated methods, such as friction-based, laser-based, and ultrasonic approaches.

Micro-joining has entered a new era in recent decades due to developments in laser technology, high-speed imaging, and micro-robotics.[2] Predictive parameter optimization and real-time quality assurance[3] are now possible thanks to the combination of computational modeling and artificial intelligence (AI)-driven process control. Furthermore, the range of material combinations that can be successfully joined has increased due to breakthroughs in surface engineering and nanostructured materials.

1.2 GENERAL REGULARITIES OF THE MICROWORLD AND LAWS OF MICRO-JOINING SCIENCE

Macro- to micro-scale transition in manufacturing is not merely an instance of scaling dimensions. The microworld is governed by special physical laws where surface forces, such as van der Waals forces and electrostatic attraction, begin dominating volumetric forces like gravity.[1] This fundamental shift has extremely important consequences for joining processes. At the microscopic level, the high surface-area-to-volume ratio means that phenomena such as surface tension, atomic diffusion, and thermal regulation become predominantly important, and phenomena based on inertia and gravitation become insignificant.

Overcoming surface problems like surface roughness, oxide formation, and contaminants to attain atomic bonding between two surfaces becomes the foundation of the science of micro-joining.[4] The main goal of any joining process is to lower the Gibbs free energy of the system by substituting lower-energy interfaces for high-energy free surfaces. In a perfect world, interatomic forces would cause two flat surfaces that are perfect to spontaneously adhere to one another. To promote this bonding, controlled energy must be applied to actual engineering surfaces, usually in the form of heat, pressure, or vibration. Thus, the precise application and management of this energy to produce a strong, dependable joint without causing damage to

❑1950 to 1970	Early micro-soldering and resistance welding
❑1980 to 1990	Introduction of laser micro-welding
❑2000	Integration with robotics and precision control
❑2010	Hybrid and multi-material joining
❑2020 to 2030	AI-driven adaptive and nano-joining system

FIGURE 1.2 Timeline of the development of micro-joining processes.

the frequently fragile and thermally sensitive micro-components is vital to the laws of micro-joining science.

1.3 SIGNIFICANCE OF MICRO-JOINING TECHNIQUES AND THEIR CLASSIFICATION

Micro-joining is very important for modern miniaturization. It includes various techniques that make permanent or temporary joints in components that range in size from a few hundred micrometers to hundreds of nanometers.[1] These techniques are important because they not only reduce size but also ensure structural integrity, provide electrical conductivity, create hermetic seals, and keep small parts working properly in complex assemblies.

These methods are essential in following domains where accuracy, reliability, and compactness are critical:[4]

1. Semiconductors and Electronics: Used for interconnects, wire bonding, and packaging in printed circuit boards (PCBs) and integrated circuits (ICs).
2. Medical Devices: Used to assemble drug delivery systems, implants, surgical instruments, and biosensors where biocompatibility is crucial.
3. Automotive and Aerospace: For producing sensors, actuators, and control systems for engine management and avionics.
4. Telecommunications: To manufacture parts for optical communication systems and mobile devices.

Micro-joining procedures can be divided into four main groups according to the physical condition of the materials at joining (Figure 1.3):

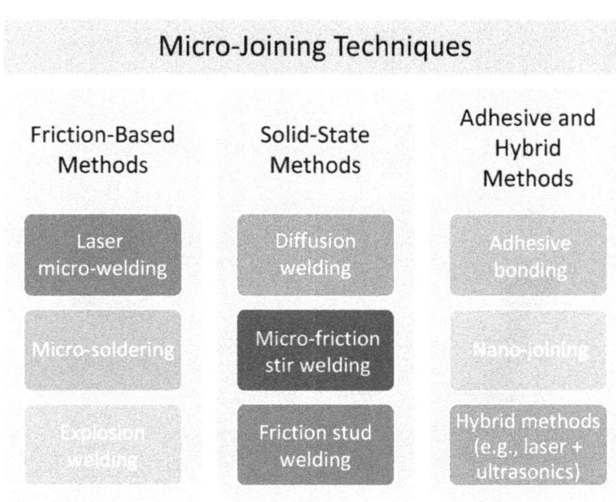

FIGURE 1.3 Classification based on the physical joining mechanism.

1. Solid-State Bonding: Methods such as friction stir welding, ultrasonic welding, and diffusion bonding that join materials without melting them.[5]
2. Fusion Welding: Methods such as laser micro-welding and electron beam welding that entail the local melting and solidification of the base materials.[6]
3. Soldering/Brazing: Methods that form the joint by using a filler metal that has a lower melting point than the base materials (e.g., reflow soldering, laser brazing).
4. Adhesive Bonding: Procedures that join components by means of chemical or physical reactions using a polymeric adhesive.

1.4 SALIENT FEATURES

A control unit, power supply, computer interface, joining tool, microscope, and micro-positioning stage are usually part of a typical micro-joining system (Figure 1.4). The control unit controls things like current, voltage, and pulse length, while the power supply gives the energy the joining process needs. Source of energy may be laser, resistance heating, or ultrasonic energy. The joining tool sends this energy directly to the workpiece, which is held securely on a micro-positioning stage that lets you make small changes in the X, Y, and Z axes. An optical system helps to line up, watch, and check the joint area at a very high level of detail. The computer interface helps to collect data, control feedback, and automate the process.

There are number of unique characteristics that distinguish micro-joining techniques from traditional joining techniques as follows:

- Accuracy and Precision: Extremely accurate positional targeting of spots as small as 10–100 μm.
- Minimal Heat Input: Many processes are made to produce heat in a very small area, known as the heat-affected zone (HAZ), which is important for components that are sensitive to heat.

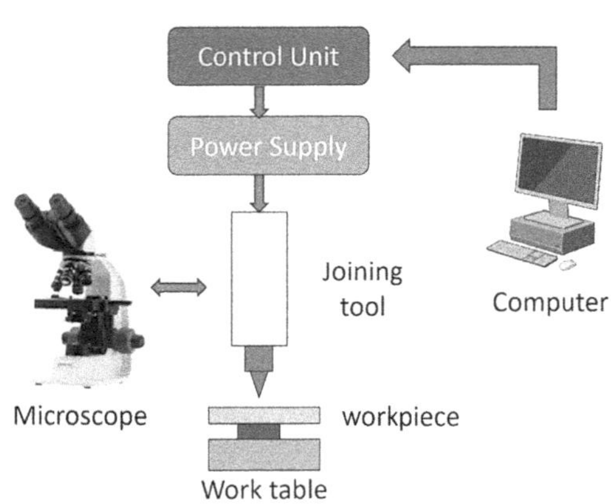

FIGURE 1.4 A typical micro-joining system.

- Low Distortion: Delicate parts maintain their dimensional integrity by minimizing warping and residual stresses due to reduced thermal input.
- Material Joining Versatility: Successfully combines disparate materials, such as metal to ceramic, polymer to metal, and different alloy combinations.
- Automation and Repeatability: These features ensure consistent joint quality and are very compatible to integration into automated, high-throughput production lines.
- Noncontact Processes: Methods such as laser welding reduce contamination and remove tool wear.

1.5 APPLICATIONS AND INDUSTRIAL RELEVANCE

Many contemporary industries rely on micro-joining technologies:

- Microelectronics and Semiconductors: For integrated circuit packaging and miniature sensors, wire bonding, flip-chip assembly, and laser welding of conductive paths are essential.[7]
- Biomedical Devices: Lab-on-a-chip systems, implants, and microsurgical instruments can be created by joining biocompatible metals, polymers, and ceramics.[6]
- Micro-joining guarantees lightweight, robust assemblies for sensors, actuators, and control systems functioning in challenging conditions in the automotive and aerospace industries.[8]
- Optoelectronics and Photonics: Submicron accuracy and low distortion are required for the precise joining of optical fibers and micro-lenses.
- Renewable Energy: Sturdy, thermally stable micro-joints are advantageous for joining energy storage systems, thin-film photovoltaic devices, and tiny fuel cells.
- Development of microelectromechanical systems[3] and micro-fluidic devices.
- Assembly of lithium-ion batteries using micro-resistance welding.

1.6 DRAWBACKS OF CONVENTIONAL JOINING

When used on micro-scale components, traditional joining techniques like arc welding have several drawbacks:

- Excessive Heat Input: Tiny components may be distorted, melted, or thermally damaged by the large heat source.[8]
- Lack of Precision: It is not appropriate for fine-pitch interconnects due to its inability to concentrate energy on a small area.
- Geometrical Restrictions: Micro-components cannot be accessed or controlled by large instruments and electrodes.
- High Residual Stress: In brittle materials, the substantial heat input may cause stresses that result in failure or cracking.
- Material Incompatibility: Dissimilar materials with different thermal properties are frequently impossible to join due to the extreme heat of conventional welding.

1.7 ADVANTAGES OF MICRO-JOINING

The following are clear benefits of micro-joining techniques:

- Preservation of Material Properties: Degradation of base material properties and metallurgical changes are avoided by low heat input.[9]
- Weight and Space Reduction: Makes it possible to create products that are smaller and lighter.
- Energy Efficiency: When compared to macro-scale processes, they consume less energy per joint.
- Improved Product Performance: Makes it feasible to produce intricate, multipurpose devices that would not be feasible to produce otherwise.
- High Strength-to-Weight Ratios: Produces joints with mechanical properties comparable to the base material.
- Eco-Friendly: Methods such as friction stir welding don't use solvents and don't release any toxic fumes.

1.8 CHALLENGES IN MICRO-JOINING

Several significant obstacles still exist in micro-joining research and practice[10]:

1. Thermal Management: To avoid dimensional distortion or microstructural deterioration, localized heat input needs to be carefully controlled.
2. Interfacial Reactions and Material Compatibility: Weak or brittle interfaces are frequently caused by mismatched thermal expansion coefficients and intermetallic formation in dissimilar materials.
3. Process monitoring and quality assurance: Sophisticated nondestructive testing (NDT) methods, like X-ray tomography and scanning acoustic microscopy, are necessary for the detection of subsurface flaws and voids.
4. Repeatability and Process Control: Because the process is sensitive to even slight changes in surface condition and environmental factors, it can be difficult to maintain consistency in micro-scale joints.
5. Sustainability and Resource Efficiency: Global trends toward environmentally sustainable manufacturing are in line with the reduction of process energy, material waste, and hazardous fluxes.

Multidisciplinary research involving physics, data science, mechanical engineering, and materials science is needed to address these issues.

1.9 COMPARISON STUDY WITH CONVENTIONAL JOINING PROCESSES

Table 1.1 provides a concise comparison between micro-joining and conventional joining processes.

To sum up, micro-joining has become a crucial and enabling field in advanced manufacturing. Innovation in a variety of high-tech industries is directly fueled by its

TABLE 1.1
Comparison of Arc Welding and Laser Welding

Parameters Considered	Conventional Joining Technique (Arc Welding)	Micro-Joining Technique (Laser Welding)
Scale	Macro (mm to m)	Micro (μm to mm)
Heat Input	High, diffuse	Low, highly localized
Heat-Affected Zone	Large	Very small, often negligible
Precision	Low	Very high in terms of μm
Implementation of Automation	Possible, but often complex	Highly probable and less complex
Dissimilar Materials	Difficult to join	Achievable
Thermal Distortion	Significant	Minimal
Typical Applications	Shipbuilding, construction, pipelines	Electronics, medical devices, microelectromechanical systems

capacity to accurately and dependably assemble tiny components. A thorough review of the various micro-joining techniques, their use with various material classes, and their influence on upcoming technological trends will be given in the upcoming chapters.

Micro-soldering is known for the complex process execution. Hence, the process needs to be addressed with precise handling and utmost care. Heating step is an integrated part of soldering process where heating of the workpiece is accomplished either through hot-air emitting gun, reflow oven, or commercial soldering iron.[5] Thermistor is a part of modern micro-soldering equipment, as shown in Figure 2.5. Thermistor is nothing but a resistor where resistance varies based on the variation in temperature. Hence, the temperature monitoring can be overviewed through thermistor. Furthermore, thermal management safeguards circuits from overheating. Highly precise, swift, and cost-effective monitoring are possible through thermistor. Rocker switch enables the micro-soldering process to be executed through power on and off. Solder station is well managed by simple operation of rocker switch. Power on/off control is user-friendly with indicating facility. The operator can recognize the indication, even handling and processing the micro-electronical components. The incorporation of the Arduino facility adds additional advantages to the soldering station. Arduino facilitates permeant, as well as reliable, joints during micro-soldering. Effective clamping of circuit board supports micro-soldering as components are joined at the micro-level. The preliminary step includes proper cleaning of the component surface where the soldering is carried out. Corrosion removal is also a part of the cleaning process. A chemical substance called flux is applied as a protection layer against corrosion and enables sufficient adhering of solder over the workpiece. The micro-soldering gun has to be heated up by turning on the heater through rocker switch. Once sufficient heating is achieved, the connecting point is applied with flux followed by solder. Flux assists and ensures the smooth flow of solder. Flux also resists oxidation in the heating stage. Micro-joining is realized when solder melts and establishes the connection. As a part of post-processing, cleaning of the component is carried out, and the inspection of micro-joints ensures the robustness of the process.

1.10 FUTURE PROSPECTS AND RESEARCH DIRECTIONS

Rapid developments in digital manufacturing, data-driven control systems, and machine learning algorithms are influencing the direction of micro-joining in the future. AI is being used more and more for autonomous joining system control, defect prediction, and real-time process optimization.

New hybrid techniques, such as laser and ultrasonics hybrid joining, seek to minimize the drawbacks of various mechanisms, increasing their benefits. In addition to that, joining procedures are being combined with technologies from microfabrication and additive manufacturing to produce smooth, multipurpose structures.

Adaptive feedback control systems, 4D process modeling, and in situ diagnostics are at the forefront to maximize accuracy and dependability. New applications in quantum devices, bio-integrated electronics, and micro-robotic systems – where joint fidelity directly affects functional performance – should be made possible by these advancements.

1.11 ARTIFICIAL INTELLIGENCE FOR AUTONOMOUS MICRO-JOINING SYSTEM

AI is being used more and more for autonomous joining system control, defect prediction, and real-time process optimization. The incorporation of AI into micro-joining processes signifies a transformative shift from conventional manufacturing methodologies to intelligent, adaptive, and self-optimizing systems. Micro-joining requires levels of accuracy, repeatability, and quality assurance that have never been seen before. AI technologies have become essential for meeting these needs. AI and micro-joining coming together solves some of the biggest problems with human-operated processes, such as optimizing parameters in real time, predicting defects, and monitoring quality at scales that humans can't see.

Figure 1.5 shows AI system architecture for laser micro-welding.

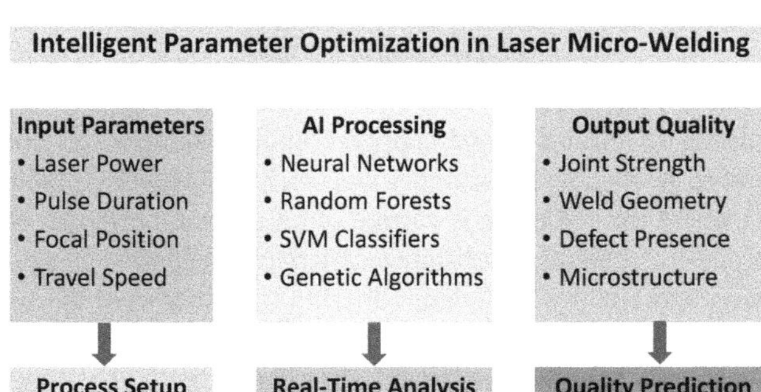

FIGURE 1.5 AI system architecture for laser micro-welding.

Using previous process data and joint quality measurement parameters, a collaborative learning model is developed that could predict the best power profiles for laser micro-welding of medical devices made of titanium grade 5 (Figure 1.5). The system had real-time thermal monitoring that changed the laser settings based on how the heat was actually spread out.[11] Reinforcement learning allowed control strategies to perform better every time based on feedback received from both sides. Traditional parameter development involved lengthy development cycles, costly materials, and a great deal of experimentation. Equipment drift and material batch variations made it challenging to maintain consistency in quality. However, by employing AI, the development time was shortened about 50%, giving a major competitive edge.

1.12 ORGANIZATION OF THE BOOK

The purpose of this volume, *Micro-Joining of Materials – Current and Future Needs*, is to give a thorough summary of the theoretical foundations, technological advancements, and novel progresses in the domain of micro-joining.

An outline of micro-joining techniques, including both conventional and contemporary approaches, is provided in Chapter 2.

The joining of metals, ceramics, polymers, glasses, and composites, is explained in Chapter 3, including joint strength issues, quality control, and sustainability issues.

The implementation of laser micro-welding in semiconductor industries, electronics industries, and biomedical systems is covered in detail in Chapter 4.

The integration of AI, automation, and nanofabrication in developing micro-joining technologies is examined in Chapter 5.

Representative case studies and industrial applications are presented in Chapter 6 to illustrate the concepts' applicability in real-world situations.

Together, these chapters provide readers with both a foundational understanding and a modern perspective on the current and future trends in micro-joining technologies.

REFERENCES

1. Zhou, Y. 2015. Micro-joining and nano-joining. *Materials Science and Engineering: R: Reports*, 90, pp. 1–35.
2. Tummala, R.R. 2001. *Fundamentals of Microsystems Packaging*. McGraw-Hill Education, pp. 1–25.
3. Esashi, M. 2008. Wafer level packaging and MEMS. *Journal of Physics: Conference Series*, 100(5), 052001.
4. Messler, R.W. 2004. Joining of materials and structures: From pragmatic process to enabling technology. *Journal of Materials Processing Technology*, 150(1–2), pp. 1–2.
5. Chen, Z., Wang, X. and Li, Y. 2020. Advanced micro-joining technologies for next-generation electronics. *Journal of Manufacturing Processes*, 56, pp. 1109–1123.
6. Stieglitz, T. 2005. Micro-joining in medical device manufacturing. *Biomedical Microdevices*, 7(2), pp. 99–103.
7. Oehmen, J. and Brombacher, W. 2002. Reliability of micro-joined assemblies. *Microelectronics Reliability*, 42(9–11), pp. 1361–1367.

8. Olah, A. and van Vliet, H.E. 2005. Laser micro-joining. *Optics and Lasers in Engineering*, 43(2), pp. 235–243.
9. Li, Y., Hu, Y. and Wang, S. 2019. Nanojoining: From fundamentals to applications. *International Materials Reviews*, 64(7), pp. 407–433.
10. Pramanik, A., Basak, A.K., Dong, Y. et al. 2017. Joining of carbon fibre reinforced polymer (CFRP) composites and aluminium alloys: A review. *Composites Part A: Applied Science and Manufacturing*, 101, pp. 1–29.
11. Zhang, C. and Chen, Y. 2022. Intelligent laser welding based on machine learning and real-time monitoring. *Optics and Laser Technology*, 156, pp. 108456.

2 Micro-Joining Methods – An Overview

2.1 INTRODUCTION

Micro-joining is the common term that represents several processes for accomplishing permanent or temporary joints at micro-level. The present technique is known for achieving structural integrity and establishing electrical conductivity for micro-components. Micro-joining also ensures the functionality of minute parts. As the components are handled at micro-level, the current process encounters several challenges during the execution. Small geometries are difficult to be handled. Micro-joining methods overcome geometrical constraints through precise handling. Most of the micro-joining methods evade the thermal distortion by handling heat under control.[1,2] Dimensional precision is achieved at microns. Thin materials can be processed effectively. The introduction of micro-joining technology has enabled manufacturing industries to produce miniature parts. Weight reduction and energy savings are facilitated by micro-joining techniques without sacrificing the product performance. Product innovations are possible through versatile micro-joining options. The current chapter provides an overview about various micro-joining methods that are widely deployed in several industrial applications.

2.2 LASER MICRO-WELDING

A wide range of tiny components are essential for producing micro-electronical and other micro-level engineering devices. Typical scale-downed parts are complex in the context of material combinations and formations. Hence, unique joining techniques are required to achieve the micro-components without compromising joining strength and quality. Furthermore, mechanically capable and sound joining processes are in demand to accomplish tiny joints with precision. Research findings explored the capabilities of laser as a source for the implementation of micro-welding for joining various means of applications. Micro-welding refers to micro-level joining and is also termed as "precision welding." Specifically, laser-based micro-welding is termed for joining less than 1 mm geometries and weldment within 1 mm depth.[3] The significant requirements of laser-based micro-joining are exhibited in Figure 2.1 for comprehensive micro-level joining. The successful laser micro-welding operations have opened way for wide variety of minute components, deployed in telecommunication industries, electronic sectors, medical fields, and automotive firms. Wide options are available in the form of laser sources for deployment in micro-welding. CO_2-powered laser, diode enabled laser, fiber-based laser, pulsed Nd:YAG laser, and disk based laser are the choices for laser selection. However, laser selection is influenced by various factors such as cycle time, material nature, tolerance, laser cost,

DOI: 10.1201/9781003502395-2

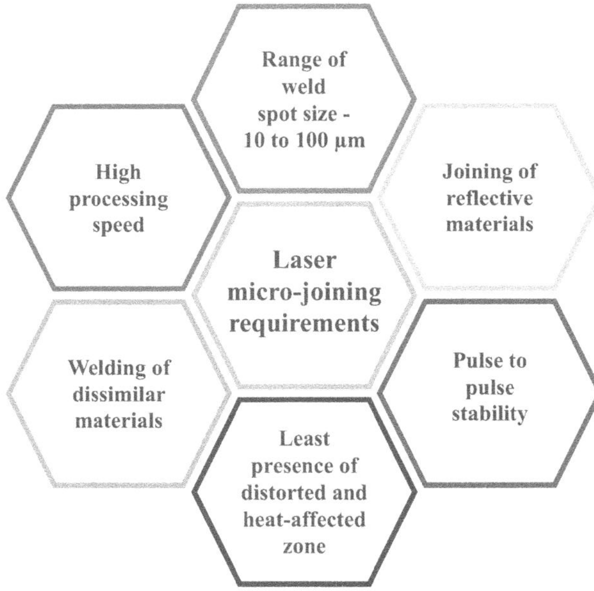

FIGURE 2.1 Significant requirements of laser-based micro-joining.

and integration needs. Especially, the cost associated with laser plays a crucial role in laser selection, as it is directly linked with manufacturing cost. Furthermore, the nature of application also influences laser selection, as the brightness of laser beam and quality varies based on the applications. Hence, high concerns are towards brightness and excellence of beam. The potential benefits of excellent beam quality can be witnessed while transporting laser through optical fiber with very small diameter. Typical laser navigation achieves minute spot size on the workpiece.

2.2.1 Execution of Laser Micro-Welding

Laser micro-welding is categorized under the fusion-based welding process. The present joining methos can be executed with or without the inclusion of filler material. Proper cleaning of the workpiece has to be ensured for maintaining the precise welding tolerances. The welding process commences with the precise positioning of the workpiece. Appropriate fixtures and jigs are deployed for the effective clamping of the workpiece. The laser beam is generated out of the laser source. For example, fiber-based lasers are activated through optical fiber. Similarly, mirrors are part of CO_2 lasers for activation. Hence, the laser source varies based on the activating medium. Furthermore, lenses are incorporated to enhance the beam's concentration through a narrow channel. The lens arrangement and propagation provide adequate energy density for melting the workpiece. Hence, the joint formation is ensured through proper melting. The molten portion should be free from external gases (oxygen and nitrogen). The presence of external gases creates welding flaws or porous weldment. The anomalies can be avoided by engaging shielding gas that

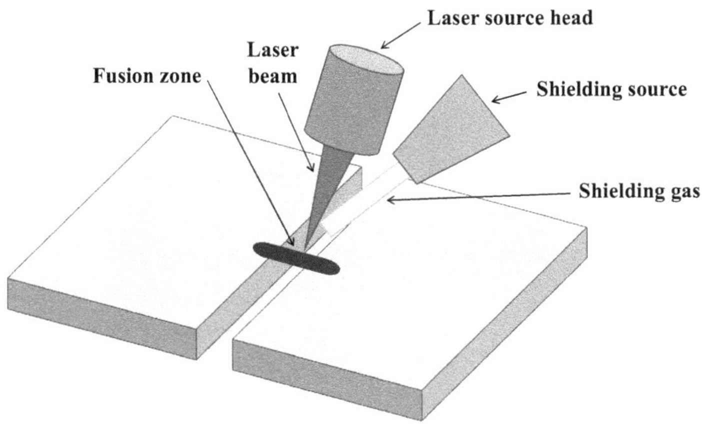

FIGURE 2.2 Laser-based micro-welding.

is supplied simultaneously, as shown in Figure 2.2. Nitrogen, argon, and helium are the renowned shielding gases. In a couple of milliseconds, the workpiece experiences enormous heat to accomplish the melting point. As the laser beam is navigated towards the direction to be welded, the molten region is solidified by the rapid cooling and ensures the sustained weld seam. A superior welding region can be accomplished through proper travel speed setting and fixing sufficient laser power. Hence, wrapping issues, residual stress formation and cracks can be avoided through optimized parameter settings. The equipment quality also plays a significant role in the determination of joint configuration. Once the welding process is over, the workpiece cools either naturally or via external cooling arrangements.

2.2.2 Merits and Demerits of Laser Micro-Welding

Laser micro-welding ensures optimal consumption of energy for achieving the welding. Aesthetically appealing weldments can be realized due to the precise joining at micro-level. However, the initial investment on the laser equipment is high and leads to expensive welding. Laser micro-welding encounters difficulties when applied over reflective materials. The detailed merits and limitations of laser micro-welding are exhibited in Figure 2.3.

2.3 MICRO-SOLDERING

The manufacturing of electronic materials is influenced by the micro-soldering process as the integration of minute electronic components aren't possible without micro-soldering. The placement of chips is enabled by the soldering process in the electronic equipment. The micro-soldering technic facilitates the assembling of minute electronic components. Furthermore, the current process ensures stable connectivity of electronic components for establishing electrical, as well as mechanical, continuity. Thus, the overall performance of components is considerably influenced by the micro-soldering process, as it establishes the integrity among various electronic

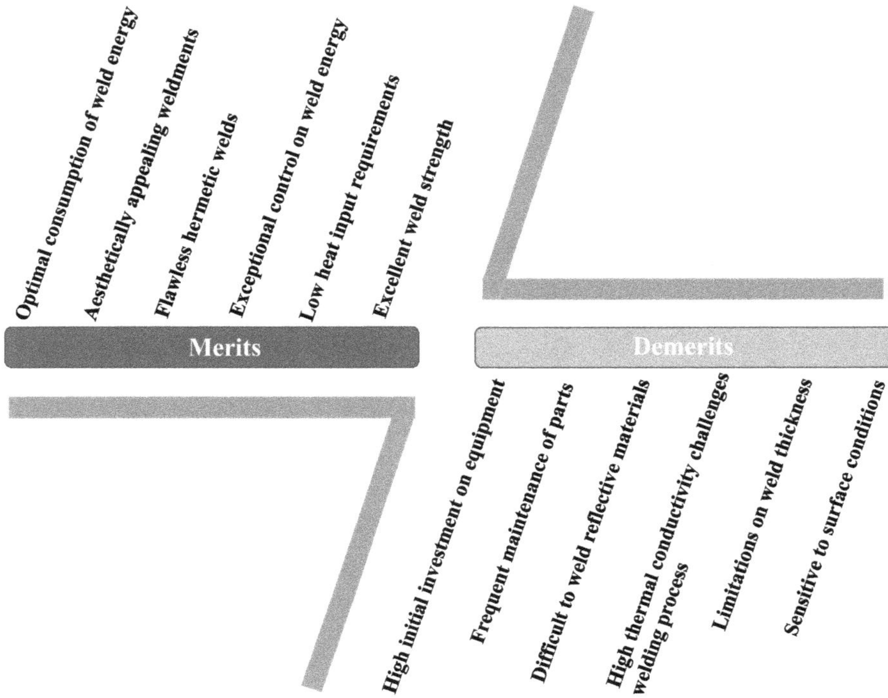

FIGURE 2.3 Merits and demerits of laser micro-welding.

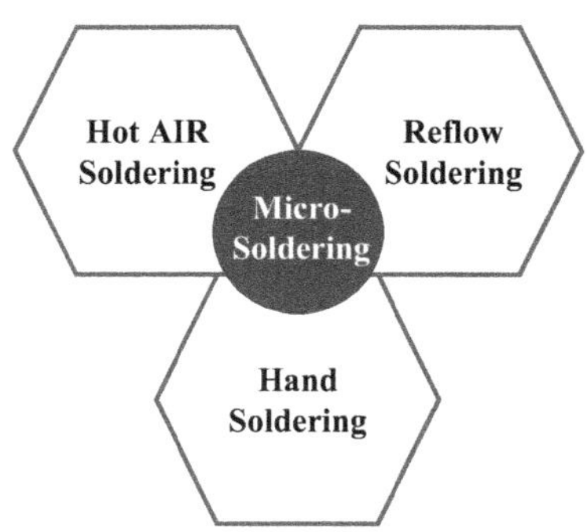

FIGURE 2.4 Classification of micro-soldering.

components in electronic equipment. The significance of the process is realized at various levels as a chip connecting medium where chips are inevitable components in electronic industries. The reliability of equipment depends on the joints made out of micro-soldering. For example, solder joints ensure the reliability of package welding. Moreover, heat dissipation is also feasible through soldered connection in the placements of semiconductors.[4] Based on the execution, micro-soldering is categorized as hot-air based, reflow type, and hand soldering (Figure 2.4). In the hot-air based soldering process, joining is achieved through the heated air that melts the material to be joined. The present technique is often suitable for larger components with relatively high execution speed. Reflow-soldering employs a heated plate for soldering components. The current soldering method holds high precision in comparison with the hot-air based method and is appropriate for smaller parts. The soldering process is deployed manually in the hand-soldering method where execution speed is relatively low. However, the present method holds the radiality in joining components.

2.3.1 Execution of Micro-Soldering

Micro-soldering is known for the complex process execution. Hence, the process needs to be addressed with precise handling and utmost care. The heating step is an integrated part of the soldering process where heating of the workpiece is accomplished either through a hot-air emitting gun, reflow oven, or commercial soldering iron.[5] Thermistor is a part of modern micro-soldering equipment, as shown in Figure 2.5. Thermistor is nothing but a resistor, where resistance varies based on the variation in temperature. Hence, the temperature monitoring can be overviewed through thermistor. Furthermore, thermal management safeguards circuits from overheating. Highly precise, swift, and cost-effective monitoring are possible through thermistor. A rocker switch enables the micro-soldering process to be executed through power on and off. A solder station is well managed by the simple operation of a rocker switch. Power on/off control is user-friendly with indicating facility. An operator can recognize the indication, even handling and processing the micro-electronical components. The incorporation of Arduino facility adds additional advantages to the soldering station. Arduino facilitates permeant, as well as reliable joints during micro-soldering. Effective clamping of circuit board supports micro-soldering as components are joined at micro-level. The preliminary step includes proper cleaning of component surface where the soldering is carried out. Corrosion removal is also a part of the cleaning process. A chemical substance called

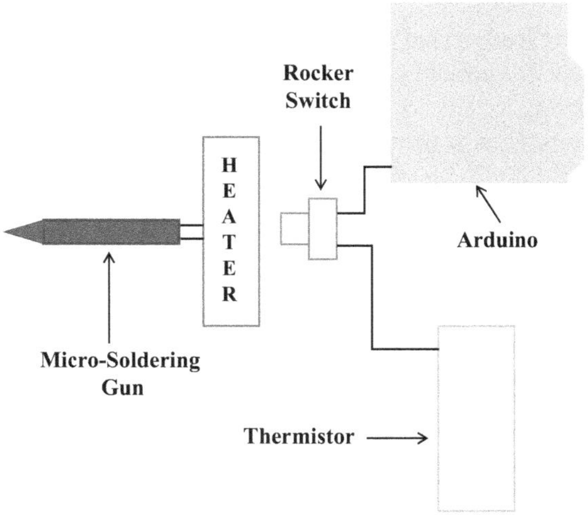

FIGURE 2.5 Micro-soldering equipment.

flux is applied as a protection layer against corrosion and enables sufficient adhering of solder over the workpiece. Micro-soldering gun has to be heated-up by turning on the heater through rocker switch. Once sufficient heating is achieved, the connecting point is applied with flux followed by solder. Flux assists and ensures smooth flow of solder. Flux also resists oxidation in the heating stage. Micro-joining is realized when solder melts and establishes the connection. As a part of post-processing, cleaning of the component is carried out, and the inspection of micro-joints ensures the robustness of process.

2.3.2 Merits of Micro-Soldering

Micro-soldering enables the repair of distinct connectors/chips without disturbing the rest of the board.

(a) Curtails expenses by precise handling of expensive parts
(b) Supports sustainability through limiting e-waste generation
(c) Allows recovery of specific data even out of damaged devices
(d) Supports restoration from devices without manufacturer's support
(e) Represents a high-skill category
(f) Generates opportunities for expertise in repair services
(g) Provides cost-effective solutions to customers by addressing the specific problems

2.3.3 Demerits of Micro-Soldering

(a) Depends on dedicated equipment for executing the process.
(b) Well-equipped skills are required for the implementation of micro-soldering.
(c) Even minute errors lead to irreparable damages.
(d) Diagnosis and repairing work consume more time.
(e) Sensitive to excess heat.
(f) Poor handling leads to the malfunctioning of components.
(g) Manufacturer's warranties cannot be claimed for repairs caused by the micro-soldering.
(h) Limited shops support services for micro-soldering based repairs.
(i) Economic viability is less for low-cost category devices.

2.4 ADHESIVE BONDING

The adhesive bonding technique is known for surface joining where two or more layers can be amalgamated together by the application of adhesives. Surfaces experience adhesive forces on the top and cohesive forces inside the adhesives. The prime motive of the creation of joints is establishing durable joints that can handle external forces without disturbing structural integrity. Adhesive bonding holds wide application spectrums, such as aerospace industries, packaging fields, construction sectors, and electronic fields. Adhesive bonding can be applied for joining plastic components, metal pieces, composite materials, and wooden parts. Adhesives are available

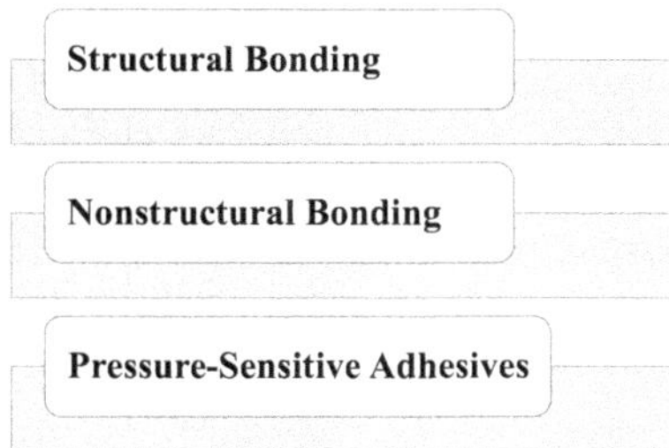

FIGURE 2.6 Types of adhesive bonding.

in paste form, liquid state, or filmy form. On solidification, adhesives establish bond between surfaces. Polyurethane polymer material, acrylic solvents, silicone paste, and epoxy are well-known adhesive bonding materials.[6] Various types of adhesive bonding are exhibited in Figure 2.6. The structural bonding category is deployed wherever highly durable joints are in need without compromising the strength. Automotive sectors and aerospace fields are effectively applying the structural bonding technique. Nonstructural bonding is preferred for temporary joint needs, nonloading conditions, and cosmetic applications. Pressure-sensitive adhesives are usually applied for sticky surfaces where less pressure is enough for creating bonds. Pressure-sensitive adhesives sustain stickiness at room temperature.

2.4.1 Execution of Adhesive Bonding

Adhesive bonding is commenced with surface preparation. Sufficient cleaning is ensured over the surface for effective bonding. Surface preparation includes the removal of dirty particles, oil substances and dusty materials. Plasma treatment, degreasing methods, chemical processing (etching) and sand application are well-known surface processing methods. Adhesive selection is also an important step in the current process. Nature of application, strength requirement, curing duration and operating conditions are essential criteria in adhesive selection.[7] Adhesives are deployed over surfaces of substrates by the brushing method, spray application, dipping technique, and extrusion. The detailed processing of adhesive bonding is shown in Figure 2.7. With the assistance of a clamping mechanism or fixtures, two surfaces of substrates are placed and aligned according to the requirement. Selected adhesives

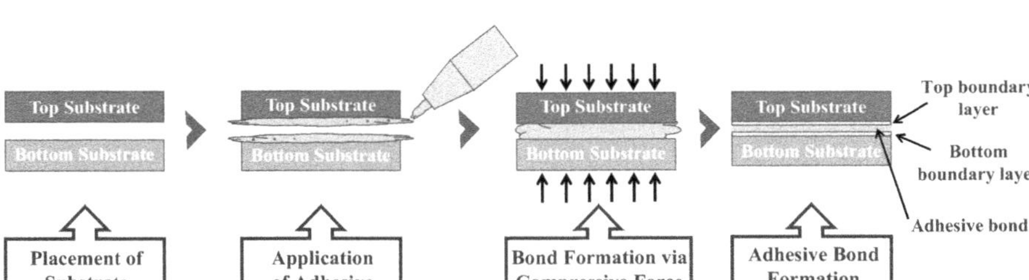

FIGURE 2.7 Execution of adhesive bonding.

are applied on both surfaces. Compressive force is applied for establishing the bond. Bonding layers consume sufficient time as a part of curing where solid bond is created. Apart from natural curing, external means of curing is accomplished through chemical application, UV sources or moisture curing. Finally, bonded workpiece is thoroughly inspected to ensure the joint quality. Visual inspection is carried out for detecting bubble formation, gaps or excess flow of materials. Furthermore, destructive testing can be utilized to evaluate durability and mechanical strength of workpiece.

2.4.2 Merits of Adhesive Bonding

(a) Stress is distributed evenly in adhesive bond.
(b) Dissimilar joints can be achieved.
(c) Curtail the weight through eliminating mechanical fasteners.
(d) Aesthetically appealing structures are possible in the absence of fastening materials and weld marks.
(e) Bonding mechanism diminishes vibrations and limits the noise generation.
(f) Effective sealing restricts the corrosion.

2.4.3 Demerits of Adhesive Bonding

(a) Poor surface preparation affects the bond formation.
(b) High temperature environment affects bonding.
(c) Limited resistance against chemical reaction.
(d) Curing process consumes much time and lowers production rate.
(e) Difficult to examine the interior quality of the bond.
(f) Limited possibilities for reworking.
(g) Restricted chances for reversibility.

2.5 MICRO-FRICTION STIR WELDING

Micro-friction stir welding (M-FSW) is a well-advanced joining technique where joints are accomplished at a solid-state condition. The current process evolved from the already well-established friction-stir welding. M-FSW supports the manufacturing industries for producing minute parts even less than 1 mm. Conventional joining techniques emit high heat during welding that may harm the minute parts of components. Hence, modern joining mediums like M-FSW are preferred where heat generation is well under control. Biomedical manufacturing companies are largely dependent on M-FSW for the accomplishment of precise joints in tiny components. Micro-manufacturing of aerospace parts can be realized through M-FSW. Thus, precision engineering is well supported by the current process. Traditional joining methods possess application constraints in the context of material limitations. The M-FSW process overcomes such material limitations faced by conventional welding techniques.[8] The presence of lightweight materials is a progressive trend and witnessed in modern-day components. M-FSW facilitates the realization of lightweight

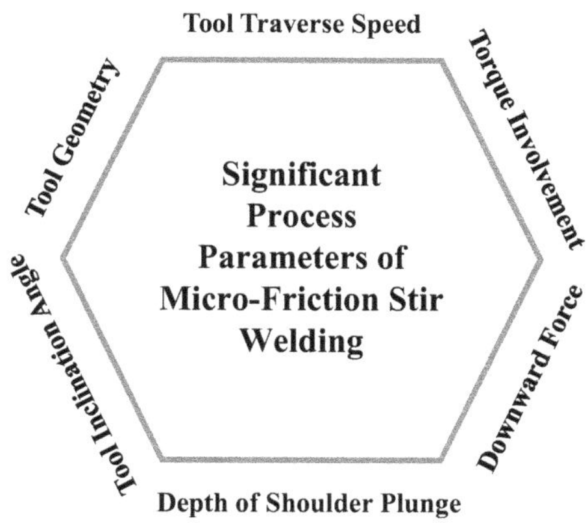

FIGURE 2.8 Significant process parameters involved in M-FSW.

materials through process support. The present welding process establishes a platform for achieving size reduction. Soft materials usually encounter processing difficulties, which can be addressed through M-FSW. Significant process parameters involved in M-FSW are shown in Figure 2.8.

2.5.1 Execution of Micro-Friction Stir Welding

M-FSW can join similar materials or components with nearby melting points. The quality of the welding can be enhanced through proper workpiece preparation. Contaminate free surface is preferred for the joining process. Oxide removal is also a part of surface preparation, achieved through acetone treatment and ultrasonic cleaning. The workpiece has to be aligned properly for precise joining. Thin sheets are engaged for clamping the minute parts to be welded. Robust weldments can be accomplished through appropriate tool selection. The nature of tool material, geometry of tool, and pin configuration are part of tool selection. The process commences when the tool approaches the workpiece and makes contact. The tool is positioned in such a way that the traveling direction follows the joining line. The tool starts rotating, and the pin penetrates into the specimen. Stirring action commences and triggers localized heat due to friction of the tool's shoulder against the workpiece. Heat generation leads to the softening of the workpiece.[9] Further penetration is achieved due to the softening of tool. As the workpiece is very thin, a higher degree of attention and control are needed over the depth of penetration. Tool rotation is sustained for the short duration to accomplish the softening of the specimen. Thus, homogeneous and even temperature distribution are ensured. The tool is navigated along the weld line with simultaneous rotation, as shown in Figure 2.9. Mechanical mixing can be witnessed due to the stirring action followed by recrystallization. Grain structures are refined with the formation of a solid joint. Once the tool reaches the end of the weld line, the tool is directed for the retraction. Natural or external cooling assists with material consolidation.

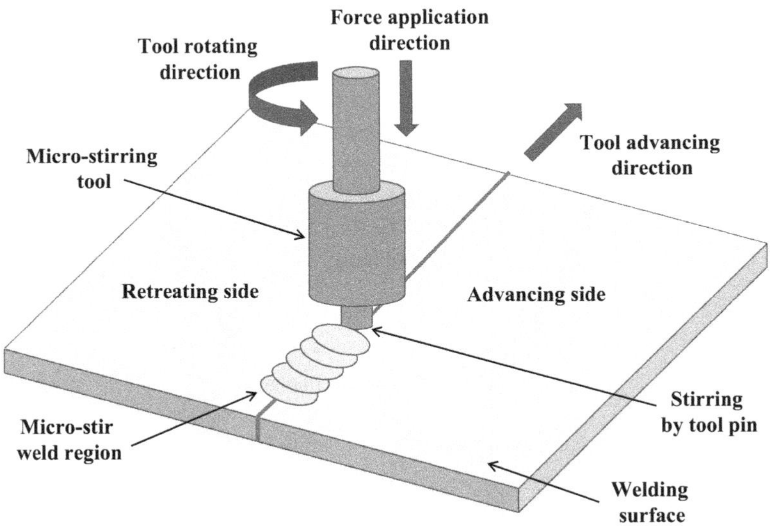

FIGURE 2.9 M-FSW.

2.5.2 Merits of Micro-Friction Stir Welding

(a) Workpiece melting is avoided. Hence, porosity related anomalies and crack defects can be avoided.
(b) Thermal distortions are evaded due to minimal heat affection.
(c) Dynamic recrystallization ensures superior weld quality.
(d) Absence of filler and shield gas make the process cost-effective and simple.
(e) Dissimilar joints are possible in thin materials as well.
(f) M-FSW can be automated.
(g) The process is known for precision and consistency.

2.5.3 Demerits of Micro-Friction Stir Welding

(a) Bond formation is disturbed due to improper surface preparation.
(b) Process has to be carried out at the environment where temperature is in a controlled condition for effective bonding.
(c) Sensitive to chemical reaction.
(d) Curing time increases the process duration.
(e) Interior quality inspection possesses complications.
(f) Rework limitations.
(g) Minimal chance for reversibility.

2.6 FRICTION-WELDING

Friction-welding is a widely recognized joining method in the solid phase. Friction is the main heat generation source. Mechanical interaction of two workpieces leads to friction and plasticizing the specimen. After accomplishing a sufficient temperature, forging pressure is applied. One of the specimens is forced against the other, and joining is achieved. The welding is executed in a couple of seconds, and joining is accomplished within a short span of time. Friction-welding can be realized in a single shot. The current process involves only workpiece materials. External materials, such as shielding gas and fluxing medium, are not needed. Smoke generation and slag formation are absent in the process. Hence, the process becomes cost-effective and eco-friendly. Friction-welding ensures the presence of fine-grained particles throughout the welding region. The quality of amalgamation is uniform across the contact. Minimal heat affection exhibits the quality of the process. The absence of solidification defects makes the joint robust. Metal dilution is also eliminated in the present process. Destructive examination reveals the consistency of welding.[10] Weld regions are strong enough, as failure is witnessed well away from the region of weldment. Strength of the material is comparable with parent material. Friction-welding is well adopted by the automotive field. Aerospace industries depend on friction-welding for material joining with dissimilar materials. Railway components can be manufactured out of the current process. The dissimilar joining capabilities of friction-welding make the process suitable for railway component manufacturing. Oil and gas industries also apply the process for similar requirements. The main categories of friction-welding are exhibited in Figure 2.10.

2.6.1 Execution of Friction-Welding

Friction-welding is initiated by thorough cleaning of workpieces to be welded. Contaminant- and corrosion-free surfaces are essential for quality weldments. One component is fixed as a nonrotating part (stationary position) with the help of a chuck (Figure 2.11). The rotating chuck mounts another component. Faces of the workpieces are to be flat for achieving proper friction and joining. The rotating part is

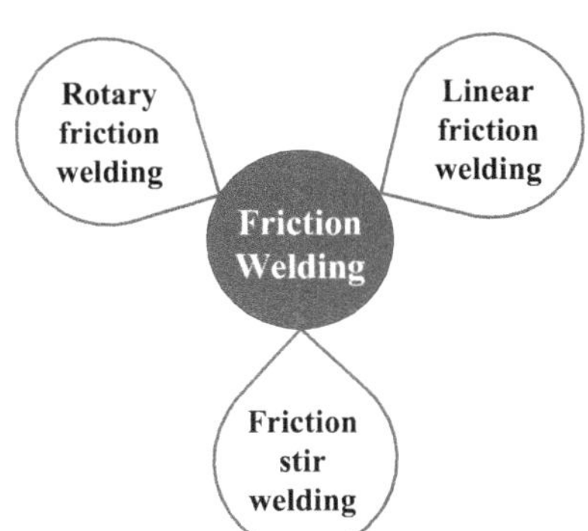

FIGURE 2.10 Major types of friction-welding.

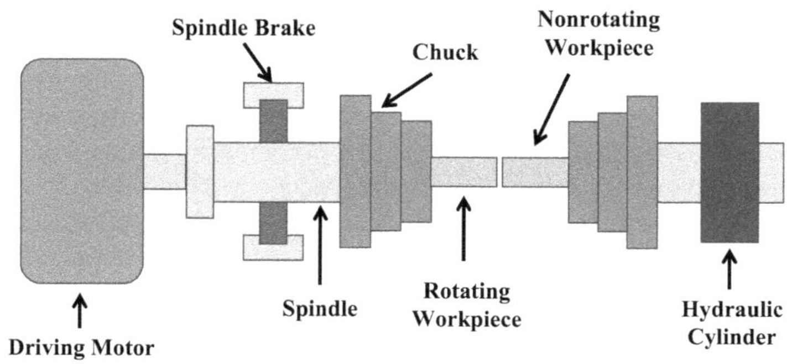

FIGURE 2.11 Friction-welding process set-up.

allowed to spin with a specific speed based on the workpiece material. Axial pressure is applied on the rotating component and navigated toward the stationary workpiece. The combination of axial force and rotating movement plasticize the material through friction generated heat. Friction is sustained for achieving the plasticization. The current state can be achieved without any exterior heat source. Under the plasticization state, materials become viscous and joining is achieved. The process is ended by stopping the rotating piece. Workpieces are allowed to cool naturally. If a flash is witnessed, proper trimming should be carried out.

2.6.2 Merits of Friction-Welding

(a) Superior weldment is possible and comparable with base material.
(b) Filler material can be excluded due to the nature of process. The same is applicable for flux materials and shielding gases.
(c) Welding of two distinct materials with dissimilar properties is possible.
(d) Heat affection is less, as materials are handled well below the melting point.
(e) Fumeless, spark-free, and emission-free execution makes the process environment friendly.
(f) More effective energy management than fusion welding.
(g) Swift operation and supports repeatability.
(h) Short span of cycle.
(i) Nature of the process supports automation.
(j) Even supports difficult-to-weld materials (example, titanium).

2.6.3 Demerits of Friction-Welding

(a) Nature of the process enables only symmetrical parts (revolving type) in case of rotary type friction-welding.
(b) Capital investment (equipment) is high.
(c) Demands the exclusion of flash as a part of post-processing.
(d) Possesses geometrical limitations, as complex shapes are difficult to join.

(e) High force is demanded by the process for workpiece clamping.
(f) Process nature demands surface preparation.

2.7 DIFFUSION BONDING

Diffusion bonding is also one of the significant processes categorized under solid-phase joining methods. The combination of pressure as well as heat enables atoms to infiltrate at the interface through diffusion. Atoms are migrated due to the high-pressure application at elevated temperature. The migration can be realized in both layers. Workpieces experience joining across the region of joint well before melting. The bonding is realized at the interfacial region and is comparable with the parent material.[11] Atomic diffusion is the prime driving mechanism behind the process. A robust metallurgical bond can be realized through the present process without dependency on filler material. Bond formation is influenced by several factors, as shown in Figure 2.12. Temperature is normally selected in the range of 50%–80% of the actual melting point. Time duration varies from minutes to a couple of hours. Oxide-free and contaminant-free surfaces are ensured in this process.

2.7.1 Execution of Diffusion Bonding

The bonding process is executed on the surface of workpieces. Thus, surface conditions are essential for the effective implementation of diffusion bonding. Flat surfaces are preferred with the absence of oxides and foreign particles. Surfaces have to be aligned properly for precise bonding. Fixtures are deployed for uniform positioning and alignment, which also ensures even distribution of pressure over the surface. Die is also preferred for organizing the workpieces. The entire assembly receives heat from furnace or heating coils as displayed in Figure 2.13. The heat process occurs in a vacuum chamber. In certain conditions, an inert gas environment is created with the assistance of argon gas. The purpose of establishing an inert/vacuum environment is to evade oxidation-related problems. Once the hydraulic pump is actuated, workpieces start experiencing the pressure under an elevated temperature environment. The current environment facilitates the exclusion of voids and other surface irregularities as the process progresses. Once the bond formation is achieved, the cooling process commences at room temperature. Distortion problems and residual stresses can be evaded through proper cooling. Bond formation is achieved in three

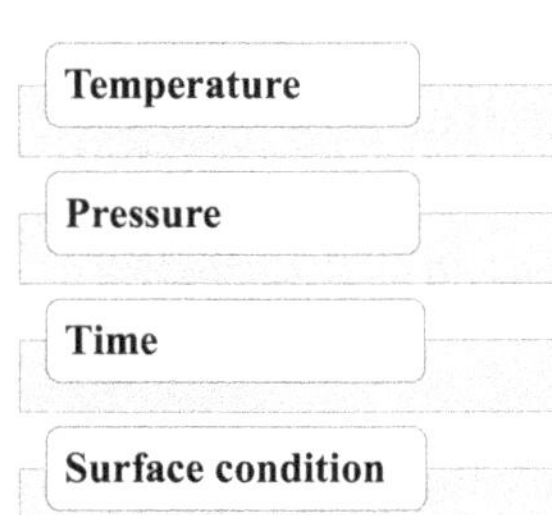

FIGURE 2.12 Key influencing factors of the bond formation.

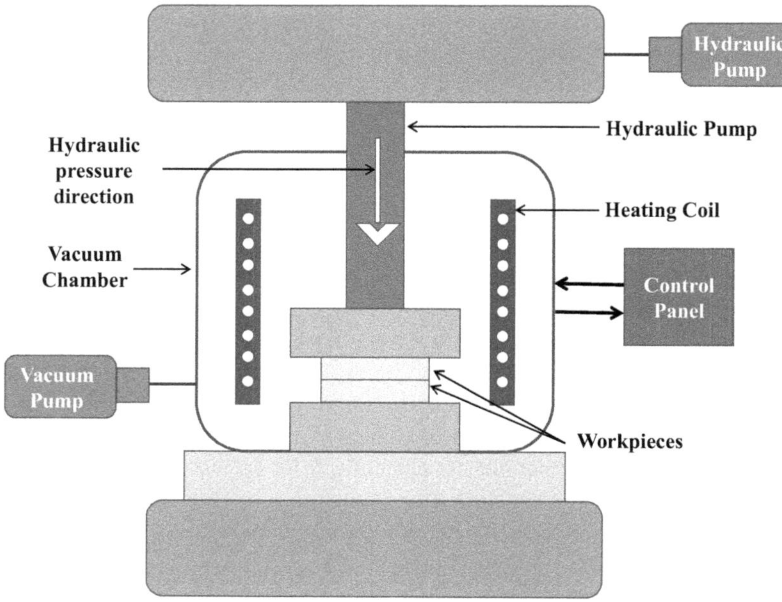

FIGURE 2.13 Diffusion bonding process set-up.

stages. Surface-level contact occurs at the initial stage. The contact points are in plastic deformation condition. Diffusion commences at the second stage, and bonding spreads across the surface. At the final or third stage, more penetration of bonding occurs. Simultaneously, voids are thoroughly eliminated.

2.7.2 Merits of Diffusion Bonding

(a) Melting medium is not required.
(b) Cost saving due to the absence of filler material.
(c) Precise dimensional control.
(d) Excellent joint strength.
(e) Thermal distortion is minimized considerably.
(f) Maximum elimination of residual stresses.
(g) Dissimilar-joint friendly process.
(h) Supports complex geometries.
(i) Possibilities for micro-level joining.

2.7.3 Demerits of Diffusion Bonding

(a) Consumes more cycle time.
(b) Poor cleaning affects the entire bond formation.
(c) Uneven surfaces lead to poor bonding.
(d) Bonding equipment is costly.

(e) Suited only for low volume application.
(f) High volume application demands process automation.
(g) Large-scale parts require exclusive set-ups.

2.8 EXPLOSION WELDING

The explosion mechanism is deployed for attaining solid-phase joints in explosion welding. Joints are achieved through energy obtained from explosive materials. The intensity of the explosives is well controlled to keep workpiece materials under the melting point. Thus, the explosive assists workpieces to attain plastic deformation. Explosives establish robust bonding at the interfacial region. Workpieces experience the metallurgical bonding. Laminated composites are well-known applications of explosion welding. The current process can be applied for achieving dissimilar joints. Explosion welding possesses significant application in the nuclear field, as the process supports the production of nuclear reactor components. The petrochemical sector and aerospace industries deploy explosion welding for component needs. Periodic waviness is witnessed at micro-level examination during the explosion that makes the workpiece anisotropic.[12] Significant process parameters of explosion welding are exhibited in Figure 2.14. A higher range of velocity is kept for the flyer-plate for achieving plastic deformation. The velocity range starts from 200 m/s and is kept up to 1000 m/s. The collision angle impacts the wave formation and affects bond quality. The angle preference range starts at 10° and is kept up to 30°. Jet formation is influenced by standoff distance. The nature of the explosive and type of loading are decided based on energy requirements of the process. Properties of workpiece materials impact the bond formation.

2.8.1 Execution of Explosion Welding

Surface preparation is the preliminary step in explosion welding. Foreign particles are thoroughly cleaned from both surfaces. If oxide formations are identified, proper

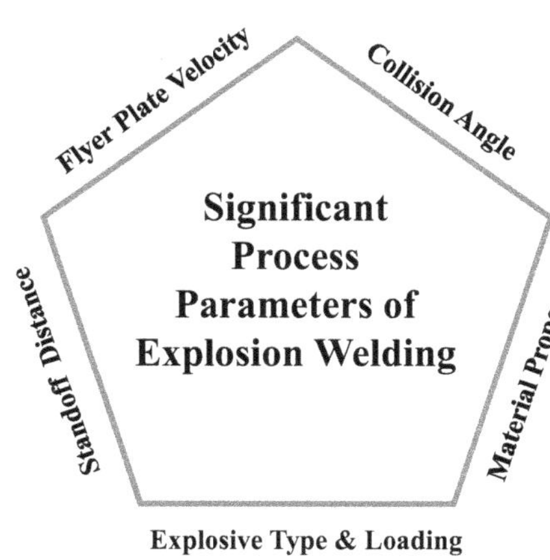

FIGURE 2.14 Key influencing parameters of explosion welding.

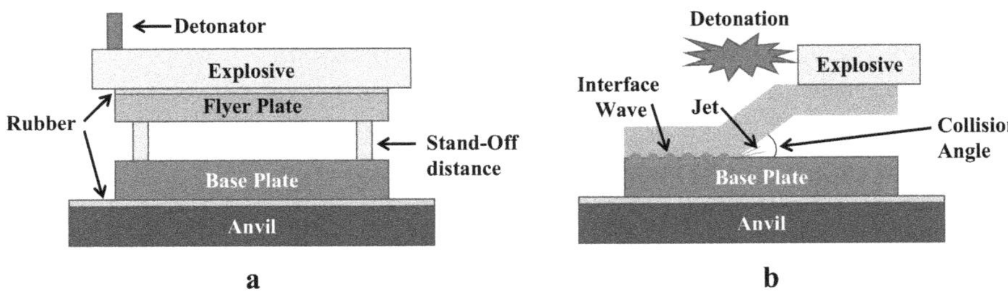

FIGURE 2.15 Process execution of explosion welding. (a) Fixation of plates and (b) welding by explosion.

cleaning is ensured. Flyer-plate and base plate are to be placed in such a way as to fix sufficient standoff distance. The fixation ensures a controlled gap between workpieces, as shown in Figure 2.15a. The upper side of the flyer-plate is occupied by explosive materials. Explosives are either evenly distributed or follow a progressive line for the detonation. The detonator is engaged to initiate the welding through explosion. The immediate effect of the explosion is experienced by the flyer-plate and pushes against the base plate, as depicted in Figure 2.15b. The smash occurs at an inclination, and the respective angle is known as the collision angle. Jetting is realized in between the plates with intensive pressure. A highly intensive collision converts material into a plastic state, and a bond is established at the interfacial region. A wavy pattern is experienced in between the bonding region, as shown in Figure 2.15b. The current process is executed in the absence of the heat-affected zone without disturbing material properties.

2.8.2 Merits of Explosion Welding

(a) Effective joining of dissimilar plates.
(b) Highly durable.
(c) Excellent reliability.
(d) Shielding gas can be avoided.
(e) Supports large-scale application in terms of surface area.
(f) Filler material is not required.
(g) Absence of thermal distortion.

2.8.3 Demerits of Explosion Welding

(a) Utmost safety is needed due to the handling of explosives.
(b) Difficult to execute on small-scale applications.
(c) Flat surfaces and cylindrical products are preferred due to geometrical limitations.
(d) Consumes more cost for initial process set-up.
(e) The nature of the process demands complex planning.

2.9 MICRO-RESISTANCE WELDING

Micro-resistance welding is an excellent selection for welding tiny parts applied in electronic components. The current process provides precise micro-joints and supports the manufacturing of intricate parts. Micro-resistance welding is categorized under resistance welding. Here, the combination of electric current and external pressure are applied for establishing the joints. Micro-resistance welding follows a similar working principle of conventional resistance-based welding. However, the process possesses modification to suit minute parts. Heat generation is achieved through resistance created by electric current. The present process demands high accuracy, precise control, and superior cleanliness for effective joining. Micro-resistance welding offers welding solutions for joints having a span of even less than a millimeter. The power consumption is comparatively lower than the traditional resistance welding.[13] The nature of the process suits joining demands for manufacturing lithium-ion-based batteries. The different types of micro-resistance welding are shown in Figure 2.16.

2.9.1 Execution of Micro-Resistance Welding

Workpiece preparation is the initial step in micro-resistance welding. A contaminant-free surface is ensured by removing foreign particles and oily materials. Ultrasonic cleaning, solvent-based surface preparation, and plasma cleaning are well-recognized cleaning methods adopted in micro-resistance welding. Workpiece compatibility has to be confirmed for both components to be welded. Also, workpiece materials should be conductive to supporting resistance-based welding. Proper placement of workpieces enables the process to achieve precise dimensions. A flat surface is highly recommended, as an uneven surface disturbs weld quality. Electrode selection is a critical step. The geometry and nature of workpiece components determines the electrode to be deployed in micro-resistance welding.[14] Tungsten is a well-known material for electrode selection. Apart from tungsten, micro-resistance welding engages molybdenum for electrode needs. The copper-tungsten combination of electrode is also suitable for the current process. The electrode is fixed on the head of the welding equipment with the help of an electrode holder, as shown in Figure 2.17. The electrode is placed in such a way that force distribution has to be uniform over the workpiece. Process parameters have to be fixed according to the welding requirements and nature of workpieces. The duration of the joining is in the range of milliseconds.

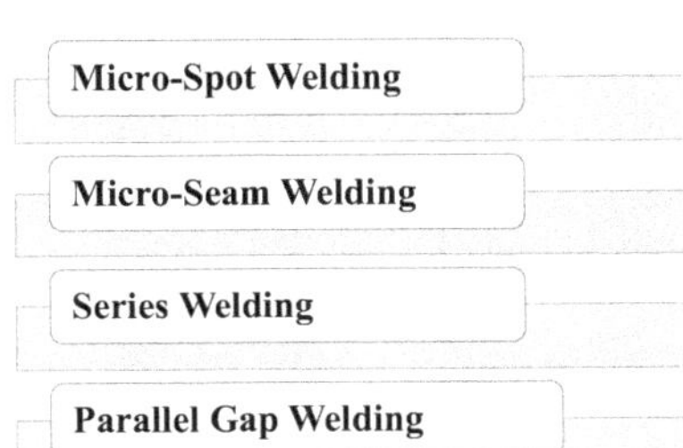

FIGURE 2.16 Types of micro-resistance welding.

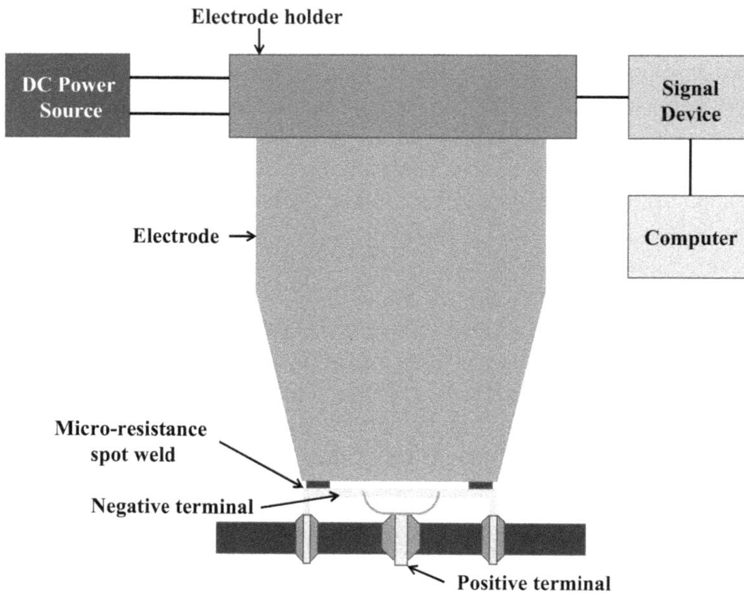

FIGURE 2.17 Process execution of micro-resistance welding.

The servo motor is engaged for the application of force. Apart from the servo motor, pneumatic-based force application is also possible in micro-resistance welding equipment. The weld cycle is commenced by manual triggering or automated mechanism. The electrode navigates to the workpieces and pressure is applied. The electric current is passed for a specific duration. The modern monitoring system overviews and records parameter values for future references. Once the joining is achieved, the weldment experiences solidification through cooling. The solidification is executed under pressure. Finally, the electrodes are retracted.

2.9.2 Merits of Micro-Resistance Welding

(a) Micro-resistance welding is known for precise execution and excellent repeatability.
(b) Well-controlled thermal distortion.
(c) Filler media can be evaded.
(d) Swift operation cycle.
(e) Supports automation.
(f) Compatible for manufacturing minute parts.

2.9.3 Demerits of Micro-Resistance Welding

(a) Compatible only with materials that conduct electricity.
(b) Complex fine-tuning of preliminary parameters.
(c) Difficult to weld dissimilar joints.

(d) Frequent maintenance is required for electrodes.
(e) Not compatible with complex geometries.

REFERENCES

1. Razul Razali, A. and Yi, Q. 2013. A review on micro-manufacturing, micro-forming and their key issues. *Procedia Engineering*, 53, 665–672. https://doi.org/10.1016/j.proeng.2013.02.086
2. Bharathi Kannan, D., Ramesh, T., and Sathiya, P. 2016. A review of similar and dissimilar micro-joining of Nitinol. *JOM*, 68, 1227–1245. https://doi.org/10.1007/s11837-016-1836-y
3. Yan, S., Zhiyong, L., Linling, S., Yingqiao, Z., and Shouzheng, W. 2023. Research and development status of laser micro-welding of aluminum-copper dissimilar metals: A review. *Optics and Lasers in Engineering*, 161, 107312. https://doi.org/10.1016/j.optlaseng.2022.107312
4. Li, L., Du, X., Chen, J. and Wu, Y. 2024. Thermal fatigue failure of micro-solder joints in electronic packaging devices: A review. *Materials*, 17(10), 2365. https://doi.org/10.3390/ma17102365
5. Xujing, N., Xue, S., Zhai, P., Luo, D. et al. 2016. Effect of Pr addition on properties of Sn-0.5Ag-0.7Cu-0.5Ga lead-free solder. *Journal of Electronic Materials*, 45, 5443–5448. https://doi.org/10.1007/s11664-016-4656-5
6. Pramanik, A., Basak, A.K., Dong, Y. et al. 2017. Joining of carbon fibre reinforced polymer (CFRP) composites and aluminium alloys: A review. *Composites Part A: Applied Science and Manufacturing*, 101, 1–29. https://doi.org/10.1016/j.compositesa.2017.06.007
7. Catarina, S., Borges, P., Alireza, A.-S., and Panayiotis, T. et al. 2023. From fundamental concepts to recent developments in the adhesive bonding technology: A general view. *Discover Mechanical Engineering*, 2, 8. https://doi.org/10.1007/s44245-023-00014-7
8. Meghnath, S., Sachindra, S., and Somnath, C. 2020. Micro-friction stir welding (μFSW): A review. *Materials Today: Proceedings*, 27, 2469–2473. https://doi.org/10.1016/j.matpr.2019.09.220
9. Sen, M. and Baran, P. A. 2025. An experimental investigation on micro-friction stir welding (μFSW) process using pinless tool. *Journal of Materials Engineering and Performance*, 34, 19652–19663. https://doi.org/10.1007/s11665-025-10635-6
10. Sathishkumar, G.B., Sethuraman, P., Chanakyan, C. et al. 2023. Friction welding of similar and dissimilar materials: A review. *Materials Today: Proceedings*, 81(2), 208–211. https://doi.org/10.1016/j.matpr.2021.03.089
11. Zhang, Z., Jie, L., Kun, L. et al. 2023. Diffusion bonding, brazing and resistance welding of zirconium alloys: A review. *Journal of Materials Research and Technology*, 26, 395–416. https://doi.org/10.1016/j.jmrt.2023.07.182
12. Zhou, Q., Honghong, L., Xuke, L. et al. 2023. Anisotropy in microstructure and shear properties of TA2/Q235 explosive welding interfaces. *Journal of Materials Research and Technology*, 25, 6472–6491, https://doi.org/10.1016/j.jmrt.2023.07.080
13. Kumar, N., Ramakrishnan, M. S., Kailasanathan, P. et al. 2022. In-depth evaluation of micro-resistance spot welding for connecting tab to 18,650 Li-ion cells for electric vehicle battery application. *The International Journal of Advanced Manufacturing Technology*, 121, 6581–6597. https://doi.org/10.1007/s00170-022-09775-z
14. Bondarenko, O. F., Bondarenko, I. V., Safronov, P. S., and Sydorets, V. M. 2013. Current and force control in micro resistance welding machines: Review and development. *Presented at International Conference-Workshop Compatibility and Power Electronics*, Ljubljana, Slovenia, 298–303. https://doi.org/10.1109/CPE.2013.6601173

3 Micro-Joining of Metals, Ceramics, Glasses, Polymers, and Composites

3.1 INTRODUCTION

Joining methods such as welding, soldering, and brazing are crucial for the creation and assembly of diverse manufactured products. These techniques encompass mechanical support and integration, electrical connectivity, optical coupling, and designed environmental safeguarding, among others. The dominant trend toward smaller-designed devices and systems, coupled with the requirement for lower costs, improved performance, and greater reliability, complicates the integration process. Micro-joining (MJ) is typically characterized as a procedure involving at least one component measuring less than several hundred micrometers yet greater than hundreds of nanometers.[1]

MJ techniques have been utilized across various industries since the 1950s. Later, a capacitor discharge machine was built for resistance welding orthodontic equipment. This concept was later implemented in the aviation industry during the 1960s.[2] In 1957, Bell Laboratories invented a method of hot-pressure welding known as thermo-compression wire bonding.[3] Ultrasonic wire bonding was invented in the 1960s, succeeded by the development of thermosonic wire bonding in 1970, which integrates both methodologies.[4] The MJ has been employed in the production of defense apparatus, aircraft, medical instruments, and microelectronics for almost 50 years. The approach has persistently faced challenges, chiefly because of ongoing developments in miniaturization. The MJ methods described in the invention have a significant impact on packaging and interconnecting in MEMS, where separate electrical, mechanical, fluidic, and optical components need to be combined and coupled with external systems. The connecting methods in MEMS are difficult to execute because of their widespread use in sensor technologies. In these conditions, they must be protected from corrosion and mechanical damage on one side while concurrently being fully exposed to the environment on the other side to guarantee accurate, unaltered physical and chemical data. Thus, MJ technology will be essential for enterprises to employ during the period of downsizing.

The first principle asserts that two ideal solid surfaces, defined by immaculate cleanliness and atomic flatness, will bond at close approach. This transpires due to interatomic forces that inherently draw them closer. The principal motivation for this merger is the decrease in the system's Gibbs free energy when high-energy free

DOI: 10.1201/9781003502395-3

surfaces are removed, reduced, or replaced by low-energy interfaces. This fundamental concept underlies all joining processes across various scales. Nonetheless, most engineering surfaces are uneven and contain oxides and other impurities that hinder the establishment of a connection. To resolve these superficial difficulties, some type of energy, usually heat and/or pressure, must be utilized. Solid-state bonding, soldering/brazing, fusion welding, and adhesive bonding are the four primary joining processes. The categories are established based on the physical state of the components. This categorization was first formulated for metals, but it later achieved extensive applicability across all joining techniques, including MJ approaches. Figure 3.1 depicts the categorization of MJ processes based on this criterion. Pressure is the principal technique for bonding solid-state materials, while diffusion inside the solid state enhances or expedites the process. Upon solidification, molten metal creates fusion welds and brazed or soldered joints. In fusion welding, the base metals melt at specific points, while in brazing and soldering, only the filler metal melts. The filler metal has melting temperatures both above and below 450°C Adhesive bonding uses

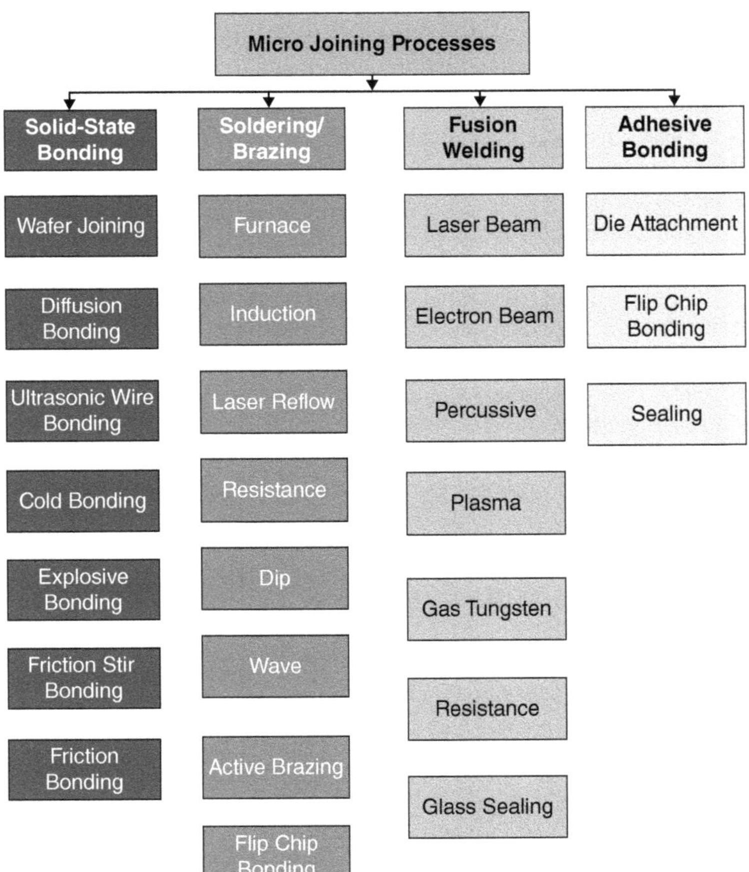

FIGURE 3.1 Classification of micro-joining process.

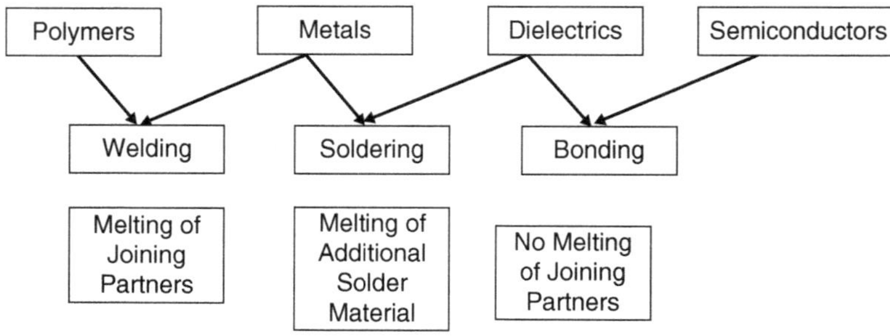

FIGURE 3.2 Classification of laser micro-joining process.

organic adhesives to connect the components. As long as the adhesives, solders, or brazes are compatible with the materials being adhered to, they can be used successfully on a variety of materials. Interlayers may be crucial in the solid-state adhesion of heterogeneous material components. The application of high-energy density beams, such as electrons or lasers, to locally melt the substrate or filler metal presents several advantages over the previously discussed methods. They can be concentrated to a few micrometers, enhancing localized temperatures while minimizing heat transfer to the workpiece. They have exceptional adaptability. Laser beam processing provides a clear advantage over electron beam processing owing to its enhanced controllability and ability to function in open air. As a result, laser MJ is becoming a feasible small-scale production technique in rapidly growing sectors like MEMS and biological applications.

Recent improvements in laser technology have facilitated integration with diverse materials employed in microsystems. Nevertheless, the efficacy of MJ mostly relies on pulse modulation methods and wavelength within a certain spectrum of laser power and intensity of laser irradiation. Figure 3.2 delineates the classification of laser MJ operations. The micro-welding method entails the amalgamation of two metals. In contrast, the processes of laser micro-soldering and laser micro-brazing integrate soldering and brazing materials, respectively, into the joining process. This section examines several facets of the micro-scale joining process, emphasizing its practical applications in real-world contexts.

3.2 SOLID-STATE BONDING PROCESS

3.2.1 Diffusion Bonding

Diffusion bonding is a solid-state adhesion method. In a hot bonding process, two components are typically fused with their surfaces in close contact at a temperature below the melting point of the base metal, facilitated by pressure. It is imperative that the fusion temperature be avoided and that filler material be rendered unneeded during the process of diffusion bonding. Diffusion bonding transpires when atoms at the interface of two adhered materials engage in diffusion. Diffusion bonding is a proficient technique for manufacturing aircraft components, since it generates a strong metallurgical bond and improves the mechanical properties of the bonded area. This method employs a foreign metal with a lower melting point to join similar or dissimilar metals, distinguishing it from brazing or the transient liquid phase (TLP) bonding procedure. No metallurgical discontinuity exists at the interface of a diffusion bond, as atoms flow across it in a solid state. The mechanical properties and microstructure of the bonded area align with those of the base metal. Titanium alloys can be effectively joined using diffusion bonding due to titanium's ability to dissolve its own oxide in a vacuum. Diffusion bonding is a solid-state technique that employs superplastic forming (SPF). It is defined as a system in which the components being united experience considerable deformation, restricted to a few percent, in the absence of a liquid phase. The methodology relies on several factors, including time, pressure, and bonding temperature, to promote atomic mobility at a microscopic scale and create a strong metallurgical link.

In some materials, the diffusion process is absent, and the application of heat and pressure creates a physical bond without significant atomic diffusion on a macroscopic scale.[1] This relates to diffusion bonding or cladding of stainless steel and copper, as well as other dissimilar metals. The combination of stainless steel and copper alloys presents challenges in achieving structurally sound diffusion bonding due to the formation of a brittle compound layer and the presence of oxygen impurities at the bond interface in the copper. An insert material is employed to enhance the binding strength. A thin layer of nickel-based alloy is often inserted between the surfaces of stainless steel and copper alloy.

The predominant technique for diffusion bonding airframes entails the utilization of thin sheets and blast shaping.[2] This method guarantees consistent pressure throughout the bonding surface, while the superplastic properties of the materials promote atomic mobility and improve diffusion. Thin-sheet diffusion bonding and blow forming may fabricate lightweight sandwich structures, which include pyramidal trusses, corrugated panels, and honeycomb panels. Massive diffusion bonding is a unique form of solid-state bonding, unlike thin-sheet diffusion bonding, which entails the fusion of two or three sheets. Massive diffusion bonding is an innovative technique for fabricating heavy-metal components by employing low inert gas pressure to consolidate many sheets in a solid state. This technique facilitates the production of components that nearly match the intended shape utilizing pre-sized titanium multi-sheets, yielding considerable weight and cost reductions. A significant advantage of massive diffusion bonding is its ability to produce heavy-metal components with reduced waste compared to alternative methods, such as mechanical machining from solid bulk materials. Closed portions can also be fabricated that are unattainable using other machining, extrusion, or forging techniques. Creep behavior is the primary mechanism for solid-state diffusion bonding.[4] Both processes initiate with heat, as thermal energy provides the requisite activation energy to surmount the potential energy barrier that limits atomic mobility.

The creep process commences with diffusion bonding, succeeded by vacancy diffusion, which eradicates all voids and promotes bonding at the joint contact.[5,6] The sealing of voids is the antithesis of the cavitation process, taking place at the end of creep rupture. The cavitation model specifies the minimal size of a stable cavity in relation to surface energy and stress.[7] In other terms, voids must reach a particular dimension before they may be completely encapsulated and adhered. The entire process is determined by surface diffusion, volume diffusion, grain boundary diffusion, and power-law creep. Certain theoretical models predict the kinetics of diffusion-related processes.[8–10]

The proposed method is to define the solid-state, diffusion-bonding model and its application in elucidating the solid-state bonding process of titanium alloys in a non-vacuum atmosphere. The complete process consists of six phases. The two surfaces must make contact in the early phase. The amount of first contact between the surfaces is contingent upon their roughness, clarity, or irregularity, in addition to the duration since the chemical etching and surface treatment were executed. Enhanced surface irregularity requires elevated pressure and extended bonding time. The second stage of diffusion bonding commences with microplastic deformation at the contact. This stage occurs when the surface ridges undergo plastic deformation, while the connecting components remain outwardly unchanged. These changes will generate gaps aligned at the interface. The apertures become distinct, and the gas pressure within them equilibrates with the pressure in the furnace. At this juncture, it is imperative to elevate the temperature. In the third step, gas absorption by the surfaces occurs, resulting in a decrease in pressure within the isolated voids. The fourth stage occurs as the voids diminish in size due to diffusion creep and the disparity in chemical potential between surface energy and volumetric energy. At this stage, the grain boundaries commence, shifting to permit contraction while maintaining a constant volume. At this point, certain little grains are diminishing in size. In the fifth stage, voids smaller than the crucial size are unstable and will rapidly vanish. A mathematical model has been employed to elucidate the mechanism of void collapse. The model considers the disparity in chemical potential between the strained interface and the unstrained surface of the void. The process that occurs when cavities form during cavitation at higher temperatures is precisely the opposite of this.[11] Ultimately, there is no discontinuity at the welded junction, which resembles the grain boundaries. The welded contact is easily discernible because of its linear arrangement. The grain boundary shifts to reduce surface energy. Ultimately, the grain boundary evolves into a seamless, continuous line to reduce surface energy.

3.2.2 Friction Stir Bonding

Friction stir welding (FSW) is a traditional solid-state joining method in which the workpieces are amalgamated at the interface through heat produced by the tool's motion and the materials' plastic deformation is shown in Figure 3.3. Friction stir welded joints do not require bulk melting, unlike traditional welding methods that include solidification procedures. FSW inherently circumvents difficulties related to solidification, including cracking, porosity, melting, and the coarsening of strengthening phases. The substantial thermomechanical deformation of FSW improves the microstructure of the weld site, potentially enhancing the weld's mechanical properties, such as tensile strength and toughness.[3] Fusion welding often undermines the integrity of the weld, but FSW can yield a weld with mechanical qualities that are comparable to or exceed those of the base metal.

This is a critical characteristic of FSW, as the weld zone produced by fusion welding often constitutes the weakest part of certain high-performance technical

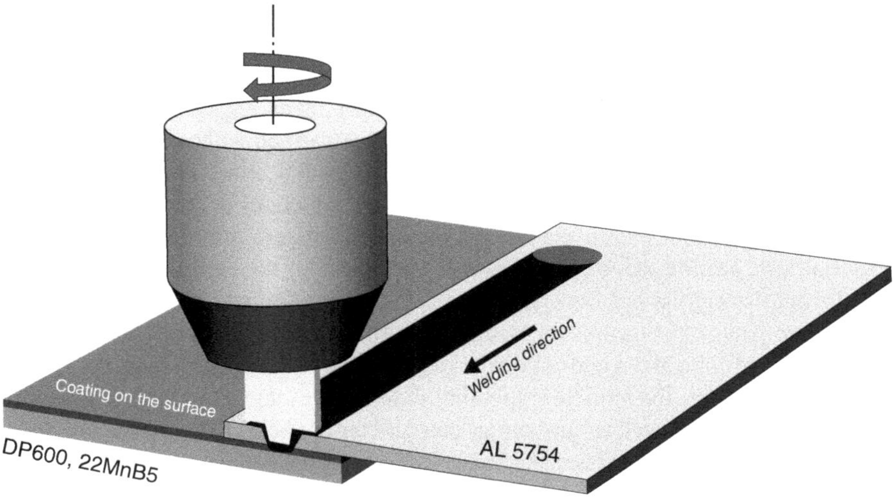

FIGURE 3.3 Friction stir bonding [4] order no-6131180468005.

materials. It has been established that strain rate, temperature, and stress fields in FSW are influenced by thermomechanical processing variables, including tool geometry, travel speed, rotation speed, and the properties of the joining materials. A variety of experimental and numerical studies, encompassing computational fluid dynamics (CFD) and the finite element method (FEM), have been undertaken to enhance comprehension of thermomechanical field properties.[4–6] A mechanistic comprehension of the interplay between bonding evolution, processing parameters, and material behavior is essential since it will eventually influence the structural integrity of the weldments.

3.2.3 Friction Bonding

Friction welding is a novel technique for uniting different materials. This method can be utilized to amalgamate analogous materials; however, it is predominantly applied to connect heterogeneous components, as traditional welding processes enable the joining of similar materials more efficiently and conveniently. It often employs two samples. One sample is fixed in a stationary collet, and the other sample is held in a spinning collet calibrated to a precise speed. After securing the specimens in their collets, the friction welding process commences, resulting in the rotating chuck spinning at the specified speed. An axial force is given to the stationary collet, causing the specimen to move toward the spinning component, as indicated by Sahin (2009). When the requisite axial pressure is exerted on the stationary collect, the two disparate materials adhere to one another. In order to achieve the ideal temperature for material softening and plastic deformation, the two specimens interact and abrade against one another. Upon attaining the burn-off length or the specified displacement, the collet halts rotation. This allows us to augment the axial pressure on the stationary component, hence improving the quality of the welded joint. The friction force, duration of friction, temperature, and displacement are important factors affecting welding. The parameters fluctuate, and the attributes of the welded specimen are employed to assess their influence.

The spindle commences spinning, and the specimens engage at the weld site under applied force. The weld interface is currently warmed without any movement. The specimens are in direct contact, leading to frictional heating. The purpose of this preliminary friction phase is to remove any superficial oxides at the weld interface.

In this subsequent phase, the size of the specimens are diminished to mitigate material loss. During the welding process, two-thirds of the entire material is relocated from its original position. This represents the preliminary phase of the operation. Three ways exist to prevent material contraction during this phase of friction. The passage of time and a constant axial distance enable their occurrence.

During this step, the spindle's spinning is forcibly stopped, and the two welded components are subjected to substantial pressure before being allowed to cool. One-third of the overall material transfer transpires during the welding process. The heat created during the second friction phase and the ultimate pressure imposed are the principal factors influencing this forging phase. The forging phase can be controlled by modifying the duration and intensity of the forging process.

3.2.4 Wafer Bonding

Wafer bonding is a process in which two mirror-polished wafers of any material adhere at ambient temperature without the application of external force or a macroscopic adhesive layer. The van der Waals force facilitates the cohesiveness of the two components. Wafer bonding at room temperature generally demonstrates lower strength than solids joined via metallic, covalent, or ionic contacts. Numerous applications necessitate the heating of room-temperature bonded wafers to augment the strength of the interfacial connections.

3.2.5 Ultrasonic Wire Bonding

Ultrasonic wire bonding is a widely used method utilized by the electronics sector to establish electrical connections in the expanding domain of power modules. The wire bonding technique comprises three primary steps: thermo-compression bonding (TC), utilizing thermal energy (300°C to 500°C); thermosonic bonding (TS), integrating thermal (100°C to 150°C) and ultrasonic energy; and ultrasonic bonding (US), relying exclusively on ultrasonic energy at ambient temperature. It is a standard technique for connecting wires to diverse electronic components, including integrated circuits (ICs), semiconductor devices, and other substrate materials. It is especially common in the production of semiconductor devices like sensors, ICs, and other microelectronic components.

3.2.6 Explosive Bonding

Explosion welding is a technique for fusing two or more metals, irrespective of their compatibility, by the significant energy released from a detonated explosion. This process creates metal composites known by a variety of names, including explosive welding metal, explosive bonding metal, and explosive welding clad. These materials are produced by numerous companies across the world. These clads are used by people all across the world in places like power plants, petrochemical companies, chemical facilities, and oil refineries.

Explosive welding clads are fabricated using significant energy generated from explosions. Metallurgical bonding facilitates the union of various similar or dissimilar metals. This is typically termed a cold-welding technique; however, the actual process resembles a metal jet.[1] A distinct gap exists between the cladding metal (flying plate) and the substrate metal (base plate). An electric detonator activates an explosive charge on the flyer plate's surface.[12] The explosion propagates swiftly, with speeds between 2000 and 3000 m/s. Upon detonation of the explosive charge, the resultant pressure at the explosion's forefront, reaching several gigapascals, propels the flyer plate at several hundred meters per second. This causes the flyer plate to touch the base plate at a precise angle, resulting in the metal being propelled forward from the point of contact.[1] The metal jet can eliminate surface oxides, adsorbed gasses, and residual contaminants on the metal surfaces. The purified surfaces are next exposed to pressures exceeding the flow strength of the metals, perhaps reaching several tens of gigapascals.[1] The atomic attractive force exerts a significant binding effect on the active metallic surfaces. Explosive welding occurs in only microseconds.

3.2.7 Cold Bonding

The cold spray (CS) process is a solid-state deposition method that utilizes compressed gas to propel microparticles to supersonic speeds. These particles cling to a substrate and/or previously applied particles to create coatings or components.[1,2] CS for thermal coating can be integrated with flame spraying; high-velocity, oxygen-fuel spraying; and plasma spraying. In CS processing, microparticles are driven at temperatures well below their melting point. This study indicates that bonding is contingent upon the kinetic energy of the particles, not thermal energy, serving as the mechanism for thermal spray processing.[3] The restriction on melting particles before impact in CS mitigates flaws arising from thermal effects, including stress, substrate and particle cracking, oxidation, and associated issues.[4,5]

CS systems are primarily classified into two groups based on the necessary propulsive gas pressure: high-pressure CS (HPCS) and low-pressure CS (LPCS). Within the HPCS framework, before injection into the nozzle throat, the gas-particle mixture is amalgamated with the hot gas. Consequently, the mixture expands via the extended diverging section of the nozzle, leading to the creation of a supersonic gas-particle stream.[13]

The compressed gas in an LPCS system, heated prior to entering the de Laval nozzle, typically demonstrates a pressure below 1 MPa.[6] The powder feeder is situated at the nozzle's diverging region. In the divergent section of the nozzle, the heated gas and powder amalgamate to create a low-pressure particle-gas stream. HPCS can accelerate particles at velocities between 800 and 1400 m/s, while LPCS is confined to launching particles at speeds ranging from 300 to 600 m/s and is limited to the deposition of lighter materials.[7] Advancement in computer science technology has resulted in the recognition and documentation of alternative techniques in the literature, such as kinetic metallization, pulsed-gas dynamic spraying, and vacuum CS.[8]

3.3 SOLDERING/BRAZING TECHNIQUES

MJ techniques allow for the careful joining of materials such as polymers, metals, ceramics, glass, and composites using specific methods designed for their characteristics as mentioned in Figure 3.4. These processes ensure strong, reliable joints in miniature and high-precision applications.[14]

3.3.1 Metals Based MJ

MJ methods, including soldering, welding, and adhesive bonding, are employed to connect small components precisely. These techniques utilize filler metals, heat, pressure, or adhesives to create strong and dependable joints, as shown in Figure 3.5.

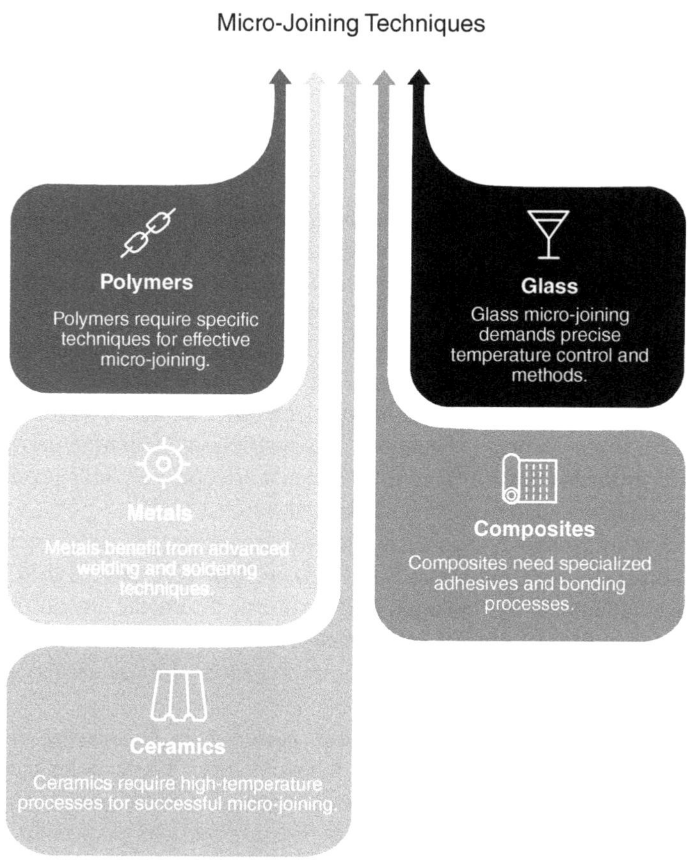

FIGURE 3.4 Soldering/brazing techniques.

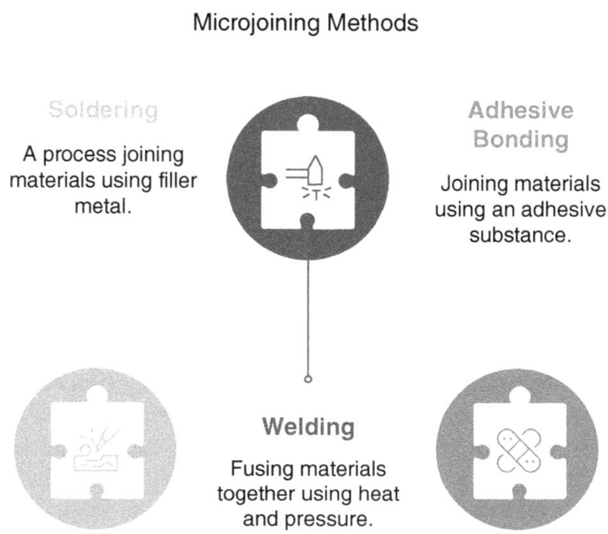

FIGURE 3.5 Metals-based micro-joining.

3.3.2 Adhesive Bonding

Adhesive bonding is a method for adhering objects in a solid state. It operates by intermolecular interactions between the components and the polymeric adhesive itself.[1] Adhesive bonding employs a polymeric adhesive that undergoes chemical or physical reactions to form a connection. The popularity of adhesive metal joining has significantly increased recently due to the emergence of strong and durable adhesives capable of withstanding both static and dynamic loads. The initiative to reduce weight in aviation, aerospace, and automotive sectors has accelerated the advancement of this joining technique, which provides distinctive weight reduction and improved stress distribution under load.[2] However, adhesive bond joints present a challenge, as they cannot be separated without causing damage. The joints are likely to deteriorate in the environment due to factors such as moisture, humidity, and temperature. However, the main obstacle that prevents individuals from using adhesive bonding is the uncertainty regarding the longevity of these joints and the difficulties involved in performing effective nondestructive testing. Environmental conditions significantly influence the longevity and ultimate mechanical efficacy of an adhesive bond joint testing. Another disadvantage is that when bonded joints are employed in engineering projects, they frequently shatter all at once rather than gradually. The surface characteristics of the components in an adhesive bond are crucial to the bonding process. Pretreating the surfaces of the workpieces prior to bonding can significantly enhance the bond strength and the durability of the connection. The surface energy of the workpieces increases during surface pretreatment, whereas the water contact angle decreases and surface tension rises. Common methods for preparing a surface for painting include solvent cleaning, altering the surface chemistry, and employing abrasion or other alterations to the surface topography. Environmental conditions significantly influence the longevity and ultimate mechanical efficacy of an adhesive bond joint. Meteorological conditions, such as temperature and humidity, are the most critical environmental factors.[3]

3.3.3 Mechanical Fastening

Mechanical fastening involves the utilization of supplementary clamping elements without welding the joint surfaces. It depends on the use of clamps or connectors, such as screws and rivets, for the eventual development of joints. Mechanical fastening properties encompass a heating cycle wherein the rivets employed for clamping undergo heating prior to fastening, allowing them to contract upon cooling, so retaining the component effectively. The heating procedure is employed exclusively for specific types of riveting; specialized mechanical activities required for the technique, including hole drilling and screw thread formation, must be conducted prior to the joining process. Various types of mechanical fastening continue to be the predominant method for attaching components owing to the process's simplicity.[4] Initially, mechanical fastening was utilized for the assembly of metallic materials (metal-to-metal), but it may now also be applied to the connecting of plastics (polymer-to-polymer) and metal to plastic. However, it has constraints, including augmented component weight and the development of tensions around fastener holes, which lead to strength deterioration and ultimately result in corrosion-related issues. Various mechanical joining techniques are available for metal to plastic connections, with a current focus on riveting because of its ability to create a reliable joint.[5] The configuration of joints frequently relies directly on service conditions, such as the necessity for leak tightness. In some cases, the joint may be engineered to accommodate variations in the coefficient of thermal expansion during assembly. Joints can be constructed to allow unrestricted movement in the plane orthogonal to the clamping member. Settineri et al.[6] examined self-piercing attachment for metal to polymer joints, revealing that the process is influenced by the physical characteristics of the rivet, including sheet thickness, tool design, and riveting force.

3.3.4 Welding

The traditional welding methods for welding different metals encompass shielded metal arc welding, gas tungsten arc welding, gas metal arc welding, and submerged arc welding. Nonetheless, the substantial energy inputs of these fusion welding methods result in poor compatibility between the materials. This issue was thicker, complicating their application in junctions between various metals and polymers. Diverse materials can be joined using innovative methods like laser welding, ultrasonic welding, friction spot welding, and FSW. Polymers such as elastomers, thermoplastics, and thermosets consist of structural macromolecules interconnected by van der Waals forces, whereas metals are composed of densely packed crystalline structures exhibiting high cohesive energy. The melting point of metals much exceeds that of polymers. Polymers typically decompose prior to the melting of metals.

3.3.5 Ultrasonic Welding

Ultrasonic welding is a solid-state welding technology that initiates combination by applying high-frequency vibrational energy and a modest clamping force. This welding technique utilizes minimal energy and necessitates that the workpieces be secured in position between the welding device (sonotrode) and an anvil using static force. Ultrasonic spot joining can employ two thin sheets or a combination of thick and thin sheets in either a basic lap joint or a butt joint, contingent upon the method of energy transmission of elastic oscillations to the welding zone. Ultrasonic vibration can be employed for welding metals and polymers; however, the methodology varies for each material, and the quality of the weld is contingent upon the application of ultrasonic energy (vibration) to the joint. In ultrasonic metal welding, the ultrasonic oscillation direction is aligned with the weld zone. During ultrasonic metal welding, the surfaces of the workpieces engage in friction, forming a solid-state bond without the need for melting. In plastic welding, the reverse occurs. The orientation of the ultrasonic oscillation is orthogonal to the welding area in ultrasonic plastic welding. Ultrasonic plastic welding liquefies and amalgamates the workpiece material at significantly lower temperatures compared to arc or laser welding techniques. Balle et al.[8] explored the process of ultrasonic metal welding aluminum sheets to thermoplastic composites that contain carbon fiber. Their study compared the efficacy of welding aluminum alloy 5754 with carbon fiber-reinforced polymer (CFRP) to other materials. The aluminum had a thickness of 1 mm, whereas the CFRP measured 2 mm in thickness. A secure and effective weld occurred at amplitudes of 40 m due to the displacement of the CFRP matrix, which improved the contact between the metal sheet and the fiber. Intermolecular interactions in the weld zone were seen when oxide layers on the metal sheet dislodged during welding, allowing the polymer matrix to be displaced from the welding zone, facilitating the ductile aluminum's adaptation to the carbon fibers. These interactions enabled the connecting partners to interlock mechanically, hence enhancing the strength of the connection. Ultimately, it was seen that the carbon fibers encasing the aluminum alloy, due to the bending of the aluminum sheet, facilitated an effective weld between the metal and the polymer.

3.3.6 Laser Welding

Laser welding is an innovative welding technique that offers new fabrication methods. It facilitates the creation and manipulation of junctions that were challenging or unfeasible with traditional welding techniques.[10] The use of laser welding of dissimilar metals has recently been studied, with a focus on welding Ti6Al4V with lead metal and welding a TiNi alloy that has shape memory properties with stainless steel, all of which are facilitated by cobalt (Co) filler metal. The joint strength of Ti6Al4V and lead metal was found to be at least as strong as the lead base metal during laser welding. Laser welding of the TiNi alloy with shape memory properties and stainless steel using Co filler metal demonstrated that optimal weld joint strength between the dissimilar metals is achievable with the appropriate thickness of Co filler metal.[15] Laser welding of dissimilar metals can occasionally produce a connection that exceeds the strength of either parent metal. Using a laser, Katayama et al.[16] studied the direct fusion of plastic and metal. An 807 nm diode laser was used to hit a 30 mm wide, 2 mm thick sheet of polyethylene terephthalate (PET) that was sitting on a 3 mm thick Type 304 stainless steel plate. Nitrogen is frequently utilized as a shielding gas to cool and clean the plastic surface. Figure 3.6 illustrates that the flow velocity was 35.1 L/min. The diode laser beam was configured to produce a line beam measuring 1.2 mm in length and 9.4 mm in width, with a laser power of 170 W and a power density of 30 W/mm^2. The laser beam velocity remained at 3 mm/s, while the shielding gas nozzle measured 0.5 mm in length and 19 mm in width. The diode laser beam could penetrate the substance (PET) approximately 90% of the time. The laser beam penetrates the transparent PET and strikes the steel surface. The laser energy elevates the temperature of the steel surface, resulting in the melting and degradation of the PET. The metal and plastic come into contact during the initial high-temperature melt because of the rapid pressure build-up from the bubble element. Tillmann et al.[17] performed a study on the optimization of processes in the laser welding of metal to polymers. The research indicated that altering the laser power from 40 W to 80 W, with other welding parameters held constant, led to a proportional enhancement in the fracture loads of the joints until a specific threshold, beyond which a decrease was noted. An increase in laser power leads to an increase in bond width, which increases the fracture load. The reported average porosities reached their zenith at the maximum laser power of 80 W. As the laser intensity decreased to approximately 70 W or 60 W, the porosity diminished, resulting in the joints achieving their maximum fracture strain. The laser power proved insufficient to completely melt the polymer component, and the steel segment was merely slightly wet, even at reduced levels of 50 W and 40 W.

3.4 POLYMER-BASED MJ

Polymer-based MJ techniques encompass ultrasonic welding, microwave welding, and adhesive bonding as shown in Figure 3.6. These methods utilize sound waves, heat, or adhesives to form strong and precise joints between polymer components.

3.4.1 Adhesive Bonding

Individuals in commerce frequently utilize adhesive bonding. The procedure entails the following steps: (1) preparing the adherend surfaces through degreasing, grinding, etching, or the application of a primer to enhance adhesion; (2) securely clamping the joining components and applying the adhesive; (3) ensuring contact between the joining surfaces while maintaining pressure; and (4) permitting the adhesive to cure and establish the joint. The curing process can take place at either elevated or ambient temperatures. Adhesive bonding has shown to be an effective technique for uniting incompatible materials. Shear is the primary mechanism via which adhesive-bonded joints transfer load. The mechanical strength of an adhesive-bonded joint is primarily influenced by the properties of the adhesive (strength and ductility), as well as the configuration of the connection. An overlap-bonded joint exhibits an uneven distribution of adhesive shear stress. The peripheries will experience the greatest stress, while the central regions will endure diminished stress. Cyanoacrylates,

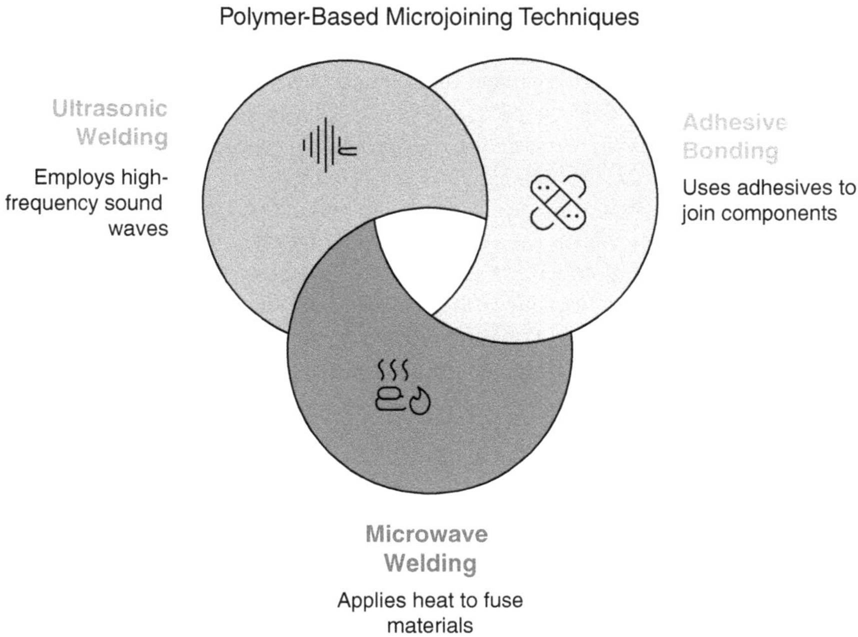

FIGURE 3.6 Polymer-based micro-joining.

silicone-based adhesives, and polyurethanes represent the predominant categories of industrial adhesives.[11]

3.4.2 Welding Processes

Established methods for welding plastics exist in various industrial sectors. Nonetheless, their application is restricted to specific categories of polymers and joint configurations. Recently, researchers have endeavored to advance welding processes applicable to a broader spectrum of materials. Advanced methodologies were established, including infrared welding, the forced mixed extrusion process, and FSW.[18] Thermoplastic composite welding remains an emerging domain within joining technology. Ageorges et al.[19] have examined the methods of connecting thermoplastic matrix composites using fusion. Almost all welding methods for plastics are applicable to thermoplastic composites, except for friction welding techniques, which can cause microstructural deterioration due to fiber fracture or realignment. Resistance welding is one of the most prevalent methods for welding composites. By analyzing their mechanical properties and microstructure, McKnight et al.[18] investigated on the resistance welding of carbon-fiber-reinforced polyetherketone composite components. Conversely, Ageorges et al.[19] examined whether glass-fiber and carbon-fiber-reinforced polyetherimide composite laminates can be welded and how process factors affect their properties. Due to the structural dissimilarity of these materials, direct welding is impractical for polymer-metal couplings. Polymers have an open structure consisting of extensive macromolecules generated by thousands of covalently linked monomer units. These units are bound by secondary interactions such as van der Waals forces.[20] In contrast, metals possess significant cohesive energy, typically two orders of magnitude greater than that of polymers, and are generally composed of densely packed crystals. This indicates that metals do not amalgamate well with polymers; rather, they typically aggregate into spherical clusters, which represent the most stable configuration under "nonwetting" conditions. The temperatures required for welding the more resilient metal with the softer polymer are typically excessive for the polymer, which often degrades prior to the metal becoming plasticized or molten.[21] Consequently, direct welding of polymer-metal multimaterial systems is unfeasible without altering the junction design. A method to achieve this involves applying a polymeric layer to the surface of the metal component and subsequently welding it to another polymeric element.[22]

3.4.3 Hot Tool Welding

Hot tooling is a widely utilized technique for welding polymers. The technique comprises the assembling of two components, which are then heated with a tool until a molten layer accumulates on their surface. The hot tool is presently retracted, and the components are subjected to continuous pressure until the molten layers solidify, forming a weld. The heating phase endures for a duration of 1 to 15 seconds, contingent upon the strength of the foundation material. For objects with substantial walls, hot tool welding cycles may exceed 60 minutes. Key considerations in hot tool welding include the tool's temperature, the heating duration, the welding pressure, the welding time, and the preparation of the components. The joining surfaces of the components must be smooth to enhance contact with the heated tool, hence facilitating improved heat transfer. To enhance the weld quality, the temperature of the heated tool must be near the polymer's melting point. The duration of welding, the pressure applied during welding, and the heating time are other factors that influence the welding process. The primary advantages of hot tool welding include (a) the absence of filler materials, (b) simplicity of assembly, (c) the capability to embed components between the joining elements, (d) compatibility with a diverse array of materials, (e) design flexibility, (f) the ability to create hermetic seals, (g) minimal emissions, (h) elevated production rates, and (i) the capacity to weld substantial components (up to 1372 mm). Hot tool welding is predominantly utilized in the construction, automotive, and pipeline sectors. Hot tool welding is employed to fabricate battery boxes, hydraulic reservoirs, and fitting/filter pipes in automobiles. PVC doors and windows welded with hot tools, as well as polyethylene pipes for gas distribution, are examples of civil engineering applications.

3.4.4 Hot Gas Welding

This welding operation utilized the traditional method of metal welding with oxy-acetylene gas, employing hot gas as the heat source. This approach involves joining two adjacent portions and placing a plastic rod along the seam. The heated gas is employed to soften the joint region and the filler rod until the polymers liquefy. Once the components have been softened, they are maintained under pressure until the joint solidifies.[20] To enhance the efficacy of the hot gas weld, it is essential to monitor the purity of the gas. Thermo-oxidative degradation may occur when heated gas contains impurities and pollutants, potentially compromising the material's integrity. Prior to welding, the joining surfaces must be cleansed with a solvent or abraded. The temperature of the gas, gas flow rate, and welding speed are all critical process parameters. The gas temperature typically ranges from 200°C to 400°C, while the flow rate varies between 15 and 60 L/min.[23] The welder's skills are the paramount determinant of their manual welding speed. The predominant materials used for hot gas welding include polypropylene (PP), polyethylene (PE), polyvinyl chloride (PVC), polycarbonate (PC), and polyamide (PA).[24] Examples of these applications include chemical storage tanks, vehicle bumper repairs, and others where the joining components are thicker than 50 mm.

3.4.5 Extrusion Welding

Hot gas welding resulted in the advancement of extrusion welding. In this process, molten polymer is extruded onto the connecting line in the form of rods or pellets. A heated gas device elevates the temperature of the joint interface to enhance adhesion. The filler material is delivered to the extruder by a wire reel, which then heats and melts it. The molten polymer is applied to the connection line and shaped into the desired bead configuration using a shoe tool. The filler material often possesses a chemical composition identical to that of the primary components. Welded parts typically possess a thickness of only 30 mm. The quality of the connection may be influenced by surface preparation, the proximity of the extruder to the weld junction, and the design and finish of the shoe-tool surface, as well as gas and extrusion temperatures, welding speed, and pressure. Extrusion welding features quicker welding cycles compared to alternative technologies, as the filler material is in a molten state upon deposition along the joint line. This reduces the number of required steps. It shares the same issues as hot gas welding due to its manual execution. Storage tanks and containers are two instances of thermoplastics that can be fused using extrusion welding. It is chosen instead of hot tool butt welding for tasks like pipelining. For conductive heat welding, there are other methods that can be used, such as forced mixed extrusion, which can be used to weld thick sections starting at 10 mm, and the bead and crevice-free technique, which is used to join pipes without producing bead or flash.

3.4.6 Microwave Welding

This approach involves placing a metal component or a conductive polymer between two welding parts and subjecting it to a microwave energy field (2.5 GHz) in a microwave oven. The insert will heat internally due to microwave energy absorption, which induces atomic and molecular vibrations. This is due to the fact that most polymers are transparent to microwave radiation. Once the molten layers are formed, the welding components are pressed together, facilitating joint consolidation. In contrast to earlier welding techniques that rely on conduction heating, microwave radiation conserves energy and reduces processing time by heating the connecting components from the interior outward.[25] This technology also possesses the advantage of facilitating internal joints, a capability not afforded by alternative joining techniques. No drawbacks have been documented in the literature to date, mostly because this is a relatively novel method. The requirement for an absorptive insert material in the joint region may be regarded as a disadvantage, as it increases the overall weight of the structure. Microwave welding is an emerging technique with limited industrial applications to date. However, individuals are seeking methods to implement it in automobiles and household devices. Current examples of materials suitable for microwave welding are extremely high molecular weight polyethylene, ABS, and polycarbonate.[26]

3.5 CERAMICS-BASED MICRO-JOINING

MJ of ceramics involves techniques such as microwave, laser, ultrasonic, and brazing methods, each offering specific benefits in precision, strength, and energy efficiency depending on material properties and application needs, as shown in Figure 3.7.

3.5.1 Brazing

Brazing is a common way to join two metal parts without melting the base metal. This process involves coalescence, which happens when the filler metal is heated to a temperature above 450°C but below the solidus temperature of the base metal.[27] In brazing, the parts that need to be joined and the metal that will fill in the gaps are heated to a temperature that will melt the filler metal without melting the base materials. The liquid filler metal must bond to the base metal surfaces and occupy the void. The filler material adheres to the connection by capillary action during cooling, therefore joining the components. The principal method employed is active

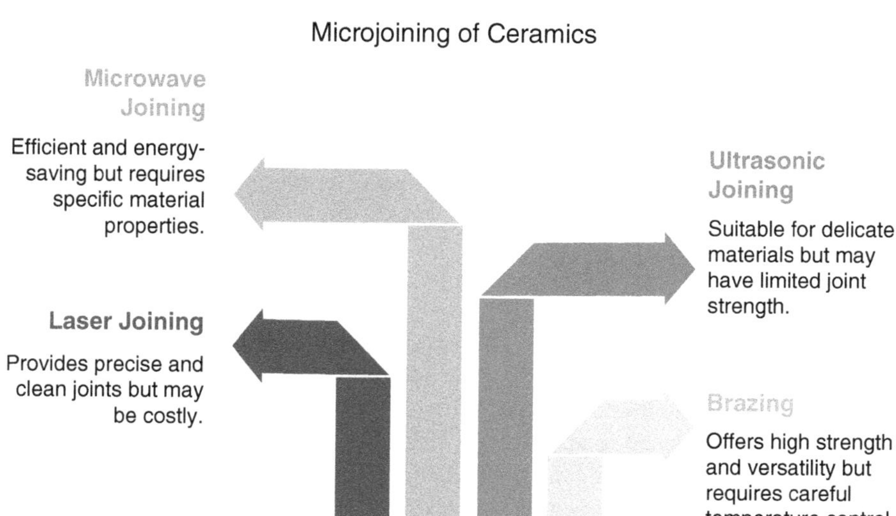

FIGURE 3.7 Micro-joining of ceramics.

metal brazing. The primary technique involved is active metal brazing. The filler alloy has reactive metals like Ti, Zr, Nb, Cr, or Y in it, which is what Al_2O_3 is made of. The active metal is added to the braze alloy to make it more wettable. This phenomenon happens when the metal reacts with the ceramic parts to make an oxide and another compound, like carbide or boron.[28] Microwave energy has been used a lot in the last few years for treating materials. Microwave radiation heats things up quickly, which shortens the time it takes to process everything. Microwave bonding with ceramics is a new area of research. Microwave brazing is used in business to make machine tools, twist drills, drill bit cutters, abrasives, ceramic substrates for electronic devices, turbocharger rotors, turbine blades, parts for car engines, nuclear fuel rods, heating elements, heat exchangers, and fuel cells.[29] A glass interlayer was used to connect the microwave energy in a home microwave oven. Meek and Blake[30] utilized a domestic microwave oven to fuse alumina to itself and to metal through a borosilicate glass interlayer. In 1988, two groups originally suggested that closed joint production without an interlayer was possible at a Materials Research Society Symposium on microwave processing materials. These groups' findings revealed that microwave heating can directly fuse alumina, mullite (3AlO.2SiO), and silicon nitride. Silberglitt et al.[31] effectively bonded SiC to SiC utilizing interlayers composed of Si, C, and Ti. Cai et al.[32] presented an innovative method for the integration of superconductors with microwave heating. Microwave-sintered Bi-Pb-Sr-Ca-Cu-O ceramic superconductors in bar form exhibited significant mechanical strength. Di Fiore et al.[33] utilized a slender interlayer to amalgamate ZnS bars by microwave hybrid heating. Arunajatesan et al.[34] utilized both EPMA and TEM to analyze the SiC/Al interfaces in microwave-bonded components. Cozzi et al.[35] amalgamated 99.5 wt. An interlayer material made up of 94% pure alumina makes up pure alumina ceramics. The interlayer material was cut into discs that were about 2 mm thick from a rod. Ahmad et al.[36] successfully bonded reaction-bonded silicon carbide without employing an interlayer or applying pressure. Sato et al.[37] integrated Al_2O_3 and MgO into a single-mode cavity microwave furnace. At the boundary between Al_2O_3 and MgO, a layer of $MgAl_2O_4$ formed. The part where the two pieces were linked was stiffer than the parts next to it. Binner et al.[38] used a microwave heating approach to attach reaction-bonded silicon carbide (RBSC) to itself and zirconia to zirconia. The whole process took around 30 minutes, with the bonding times not going over 10 minutes. Al-Assafi et al.[39] employed alumina gel as an interlayer and microwave energy as the heating source to bond alumina ceramics with differing purities. Direct joining and gel joining were executed, followed by comparative assessments of the two joint types. Bruce et al.[40] used silicon carbide composites reinforced with silicon carbide fibers for fusion applications. The composites were bonded in a microwave heating environment utilizing preceramic polymer slurry as the adhesive. Kondo et al.[41] created a connection between silicon nitride by placing a glass junction between two silicon nitride tubes and covering the joint with a silicon carbide susceptor. In a certain area, the microwave radiation made the susceptor hotter. It is necessary to heat materials evenly while they are being worked on. In ceramic sintering, uneven temperature distributions can have detrimental effects.[42] Two strategies have been used to correct this kind of unevenness. One method uses an insulator to cover the heated material, and the other uses a blend of heating processes. There have been two suggestions for hybrid heating techniques. Microwave hybrid heating (MHH) is a type of heating that uses susceptors, which are materials that can absorb microwave energy well within a certain temperature range, to add heat to the sample via conduction and radiation. At room temperature, alumina does not absorb microwaves well. As the temperature rises, alumina can take in a lot of microwave radiation. The SiC susceptors around the alumina cause its first temperature rise via heat conduction and radiation. These susceptors can take in microwave energy even when it's cold. Alumina ceramics quickly absorb microwave radiation when heated to 800°C. Researchers frequently employ a cavity featuring many modes and MHH. There is a lot of interest in using silicon carbide for microwave connections since it is a powerful microwave absorber. The microwave absorption properties of RBSC and hot-pressed silicon carbide ceramics are distinct. Silicon metal remains in the microstructure of RBSC. The MMH technique warms things up well, whether they are see-through or not. Adding an insulator can help heat spread evenly. There are two problems with the MMH approach that need to be pointed out. Thermal conduction makes it difficult for microwave energy to quickly heat objects up at first. Second, microwave radiation doesn't quickly penetrate into heated materials at some points in the procedure.[43]

3.5.2 Ultrasonic Welding

Ultrasonic welding utilizes high-frequency ultrasonic acoustic vibrations to join materials under pressure, producing a solid-state weld. The system encounters two categories of forces: a static normal force exerted perpendicularly to the interface between the workpieces and an oscillating force produced by the sonotrode's oscillation parallel to the material interface surface. The interplay of static and oscillatory stresses induces deformation that facilitates welding. Ultrasonic welding can be performed through many methods, including direct welding of ceramics and metals, vacuum deposition of metal coatings on ceramic surfaces, and insertion of activated metals between surfaces. The characteristics of the interface significantly influence the efficacy of welding between metals and ceramics. Prior studies have demonstrated that each metal-ceramic pairing possesses optimal welding parameters. Welding performed under optimal conditions significantly enhances the strength and other characteristics of the joint.[44] A significant advantage of this welding technique is its minimal heat generation, which preserves the material's strength. It can be accomplished in any environment, whether aerial or in a vacuum.

3.5.3 Laser-Based Micro-Joining

Laser-induced brazing may function as a technique to unite several ceramic components of a thermoelectric generator (TEG) unit, encompassing n-type and p-type ceramic bars, as well as a ceramic substrate. TiOx and BxC were employed as thermoelectric components to fabricate the TEGs, while AlN functioned as the substrate material. The time needed for joining is considerably shorter than that of traditional furnace brazing techniques. Moreover, it facilitates the use of a single filler system for all contact points of the thermoelectric unit. Thermoelectric modules provide diverse applications. An example represents the aggregation of energy to operate devices with minimal energy consumption. Recent investigations suggest that thermoelectric modules can be utilized in procedures that transform waste heat into usable electrical energy. Thermoelectric modules can link thermoelectric materials to the heat generated by solar collectors.[45] We can group the thermoelectric materials they have used so far by the temperatures at which they work or the kinds of materials they are constructed from. Group V elements (Bi, Sb) and their alloys exhibit optimal performance at low temperatures. Bismuth tellurides (Bi_2Te_3) and their solid solutions, consisting of Bi_2Te_3, Bi_2Se_3, and Sb_2Te_3, demonstrate peak efficacy at room temperature. These materials are presently accessible on the market; however, their usage is restricted to temperatures not surpassing 250°C. Lead tellurides, rare earth oxides, and SiGe and $FeSi_2$ alloys can be used at temperatures close to 700°C. In light of environmental concerns, the use of lead-containing components should be reduced or completely replaced with more sustainable alternatives. The thermal conductivity of titanium suboxides varies between 1 and 3 W/(m K), which is considered advantageous. The aim is to amalgamate several suboxides to optimize the ZT (figure of merit). Boron carbide ceramics are recognized as p-type candidates for high-temperature thermoelectric energy conversion; specifically, boron-rich carbides ($B_{41}xC_0$, x, 6.4) exhibit favorable thermoelectric characteristics, including a substantial Seebeck coefficient, moderate electrical conductivity, and comparatively low thermal conductivity. The exceptional thermoelectric characteristics of boron carbides are attributed to their distinctive crystal structure. Boron carbides manifest as a singular phase and display a wide range of homogeneity, varying from 8.8 at.% carbon (B10.4C) to 20 at.% carbon (B4C). Their basic cells are rhombohedral, consisting of 12-atom icosahedra and 3-atom chains. Wood recorded a significant ZT value of 0.65 for boron carbide with 10 at.% carbon and 0.5 at.% magnesium at 1000°C. A major barrier to its extensive application has been the failure to discover a viable n-type boron carbide. Thermoelectric applications often require appropriate p-type and n-type legs. Due to its high thermal expansion coefficient and great chemical stability at high temperatures, titanium suboxide is a good n-type replacement for boron carbide. Several articles indicate that the ZT values of TiOx are approximately 0.2. Currently, the focus is on materials that have better thermoelectric capabilities and the best ways to connect and integrate them. The essential factor is that the components of the TEG module must be interconnected with robust, low-resistance connections capable of conducting both electricity and heat. These connectors must exhibit minimum electrical resistance to prevent detrimental effects on the module's overall performance. Sheets of copper, silver, or molybdenum have been employed to date. Silver-containing pastes with n-type powder or resin additions have been employed to create connections between the n-type and p-type components. Experiments investigating contact formation without silver components were unsuccessful, although the copper electrode melted at the junction temperature.[46]

3.6 GLASSES-BASED MICRO-JOINING

The glass offers several beneficial characteristics, notably high compressive strength, corrosion resistance, and excellent optical features. However, glass isn't suitable for

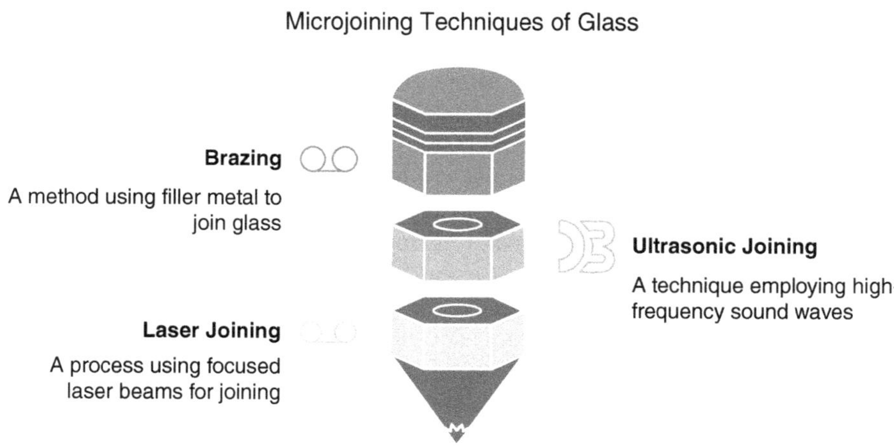

FIGURE 3.8 Glasses-based micro-joining.

many things because it fractures easily and doesn't withstand impact effectively, which is mentioned in Figure 3.8.

3.6.1 Brazing

Glass is an inorganic, nonmetallic substance that doesn't have a shape. It is challenging to adhere glass to metal since glass does not hold water well. There are two primary methods to enhance the wettability of the glass interface: active brazing[24] and ultrasonic-assisted brazing.[47] Incorporating active elements (Sr, Zr, Ti) into the filler metal can enhance wettability. These materials disperse across the glass surface, creating a reactive layer adjacent to the glass contact and facilitating a stable connection.[48] Ultrasonic waves introduced into liquid filler metal during ultrasonic-assisted brazing induce cavitation and acoustic flow. During a specific acoustic interval, the cavitation effect induces the formation, expansion, and subsequent rupture of micro-bubbles. The liquid/solid interface facilitates the formation of micro-bubbles due to the alterations in sound waves that induce tensile and compressive stress. The implosion of micro-bubbles during the acoustic phase may generate shockwaves and microjets, resulting in a high-pressure and high-temperature environment on the surface of the substrate materials. The microjet and shockwaves would erode the solid's interface, facilitating the wetting of the solid by liquids.[49,50]

3.6.2 Active Brazing

The current study examines the microstructure and properties of the interface between fluorosilicate glass-ceramic and titanium, composed of 71% silver, 27% copper, and 2% titanium by weight. The temperatures for brazing were between 850 and 950 degrees Celsius. It was challenging to figure out how to respond to the titanium and glass-ceramic. The layer that reacted on the glass-ceramic was around 4 to 5 μm thick. The reaction layer is made up of 80% titanium and 15% silicon, which means that the glass-ceramic and the brazing alloy reacted with each other. The Cu atom migrated to the Ti side, which created the complex Ti_2Cu. The chemical reaction is the most significant part of a good relationship. This was the first time that glass-ceramic made of fluorosilicate was brazed. Because of this, new methods for joining glass and metal will be developed in the future. They looked at the parts and microstructure of the joint, but they didn't assess how well the pieces fit together or how strong the junction was. The microstructure and shear strength of the joint can alter when you braze. A pure titanium/fluorosilicate glass-ceramic connection was made with different quantities of Ag-Cu-Ti alloy.[51–57] The joint's microstructure was examined, revealing the presence of nine response layers inside the joint as a whole. The brazing alloy made some reaction products when it touched the glass-ceramic. A layer of reaction developed on the surface of the glass-ceramic. The procedure of joining worked well because reaction products and a reaction layer formed. They indicated that the temperature at which the brazing took place affected the reaction layer's development, but not the titanium. The brazing temperature did not have a big effect on the different types of reaction products, albeit.[58] A vacuum is employed throughout the brazing process, and the temperature is adjusted to produce the best shear strength. The levels of brittle intermetallics likewise go greater when the temperature rises up. The brittle intermetallic compounds would make the joints weaker, which would be a huge problem. The equipment had to work harder because the brazing temperature was higher and the air was a vacuum.[59]

3.6.3 Ultrasonic-Assisted Brazing

The brazing process can be improved by removing the oxidation layer from materials and fine-tuning the grain structure with ultrasonic vibration. Zhang et al.[60] investigated the application of ultrasonic waves to improve the brazing of Zr-based bulk metallic glass with Zn-3Al alloy. Post-ultrasonic brazing, the metallic glass junction exhibited a robust adhesion. The combined effects of sonic cavitation and interfacial reactions may alter the microscopic appearance of the Zn-3Al/metallic glass interface. With an increase in ultrasonic vibration duration, the shear strength initially ascended and then later diminished. At the joining contact, two varieties of wavy-structured sublayers were present when the ultrasonic vibration duration was 96 seconds. One sublayer is black, while the other is gray. The black sublayer consisted of $Zn_{50}Al_{25}Zr_{25}$, and the gray sublayer comprised $Zn_{22}Zr$. The maximum shear strength attained was 100 MPa. Zhang et al. examined the effect of ultrasonic vibration on the microstructure and mechanical characteristics of joints. We also discussed the union of metallic glass and filler metal, which enhanced our understanding of ultrasonic-assisted brazing. Cui et al.[61] investigated the brazing of sapphire glass and metal. The coefficient of thermal expansion (CTE) between sapphire and metal was aligned to prevent junction cracking. Ultrasonic assistance was employed to apply a Zn-Al alloy coating on the contact surface of sapphire glass. The ultrasonic brazing technique produced a connection composed of a novel composite material known as SiCp/Zn-Al-Si. The maximum shear strength was 155 MPa, exceeding that of the joint made with Zn-Al alloy. Ultrasonic-assisted brazing is superior to active brazing under low-temperature and atmospheric circumstances. Ultrasonic waves can enhance the flow of filler metal and strengthen the adhesion between brazing materials.[62,63] Ultrasonic brazing is an effective method for adhering glass, hence expanding its applications.

3.6.4 Laser Welding

Laser technology is employed in microtechnology due to its little heat input, exceptional flexibility, and high efficacy. Laser welding exists in two forms: pulsed laser and continuous laser. Pulsed laser welding is mostly employed for joining small components and is particularly effective for precision equipment. Nevertheless, continuous laser welding mostly employs a gas laser for the welding of sheet metal, with carbon dioxide as the medium.[64] The interaction of ultrashort pulse lasers with transparent materials such as glass results in minimal heat effects and nonlinear absorption, rendering it ideal for micro-welding transparent substances. Currently, it has been successfully utilized for high-precision welding of microparts, such as electronic chips, which has emerged as a focal point of research in the domain of glass micro-machining.[65] Initially, the glass and metal made contact. The short-pulsed laser penetrated the glass and struck the metal surface. Employing a short-pulsed laser to fuse materials maintained their temperature at a low level. Excessive irradiation duration and elevated energy levels on the glass surface may result in pitting and cracking. When the intensity of laser irradiation is below 400 μJ, the shear strength increases. However, the shear strength remained negligible when the energy reached 500 μJ. The experimental phenomenon was clear, but the theoretical analysis was not thorough. We must acquire further knowledge regarding laser welding. Laser soldering is extensively utilized in microelectronics and other diminutive components due to its exceptional attributes. Laser welding may be the optimal selection when a low bonding temperature and a minimal bonding area are required. The concept governing laser welding between glass and metal restricts the materials eligible for welding. The CTE of welding materials is similar; thus, borosilicate glass represents a category of glass, whereas Fe-based alloy constitutes a category of metal. Advancements in smart technology are leading to an increase in the production of laser devices. In the future, laser welding will evolve to be more sophisticated. Advanced laser devices enhance the efficacy of laser welding.

3.6.5 Laser-Assisted Bonding

A laser with a high power and pulse rate can permanently bond two glass pieces without the use of an intervening layer. However, when employing a high-power laser, the entire irradiated region undergoes melting, hence impacting the material.[66] When a high-repetition-rate laser is pointed at the interior of a clear material, the time between laser pulses becomes less than the time needed for the absorbed radiation to diffuse. As a result, heat builds up in the focus volume, which makes the material's absorption coefficient a high-order function of the intensity of the incoming light. This is referred to as nonlinear absorption.[67] Exceeding the melting point will irrevocably alter the dynamics of resolidification within the focal volume, affecting its density and consequently its refractive index. This action occurs just at the interface, enabling the connection of materials with high CTE without inducing cracks. This technology utilizes high-repetition-rate pulsed lasers to facilitate adhesion between glass surfaces. The melting of the focus volume occurs by chemical bonding, which is the mechanism underlying this form of laser-assisted melting. For an effective hermetic seal, the laser's focal point must be precisely positioned beneath the interface of the bonding surfaces, while the dimensions of the heat-affected zone, repetition rate, and scanning speed must be meticulously selected to ensure the integrity of the seal.[68] Chen et al.[69] presented a thorough analysis of the laser-assisted bonding absorption mechanism in silica glass. A plasma zone develops above the focus point as heat accumulates from absorption. This area is encircled by a thermally impacted zone. The plasma and heat-affected zone evolve over time (with the number of pulses) at a pulse energy of 21.25 μJ, a repetition rate of 0.4 MHz, and a scanning speed of 2 mm/s. Nonlinear absorption results in the first pulses being absorbed near the geometric focus. As the quantity of pulses increases, the temperature rises, resulting in an expansion of the heat-affected zone laterally. The formation of a bond occurs when the material in the plasma region intermingles across the interface during this process. To attain effective laser-assisted bonding, the peak intensity of the focused laser beam must be adequately elevated to produce free electrons through nonlinear absorption. The liberated electrons later act as seed electrons for plasma generation, avalanche ionization, and single-photon absorption.[70] Richter et al.[71] successfully fused glasses with varying CTE, specifically borosilicate glass to silica glass and ultra-low expansion (ULE) glass to silica glass. The authors found that, depending on the laser settings, there exists an optimal gap between pulses that enhances binding strength. A reduced separation length increases the size of the molten zone. However, abbreviated separation lengths also induce heightened stress. Certain authors assert that intervening layers can enhance bonding; nevertheless, they are not essential for laser-assisted direct bonding. De Pablos-Martín et al.[72] introduced a 50 nm titanium layer between two borosilicate glass substrates, while Carvalho et al.[73] integrated a biotin monolayer together with a 100 nm chromium layer between two borosilicate glass substrates.

3.7 COMPOSITES-BASED MJ

3.7.1 Ultrasonic Welding

Ultrasonic welding is the predominant method for welding thermoplastics in Figure 3.9. It is utilized in the production of floppy disks, medical apparatus, battery casings, and diverse automotive components. Ultrasonic welding entails applying pressure to the components to be fused while subjecting them to ultrasonic vibrations that are perpendicular to the interface. Thermal energy is produced in the material by elevated frequency stresses, and if the different parts are fabricated properly, it can be generated at the joint interface by a synergy of friction and hysteresis. Ultrasonic welding can occur at varying distances from the horn, ranging from a few millimeters to many centimeters. The control process exhibits heightened sensitivity at extended distances. The usefulness of flexible polymers is limited by their energy absorption; however, this concern does not apply to their composites. In ultrasonic

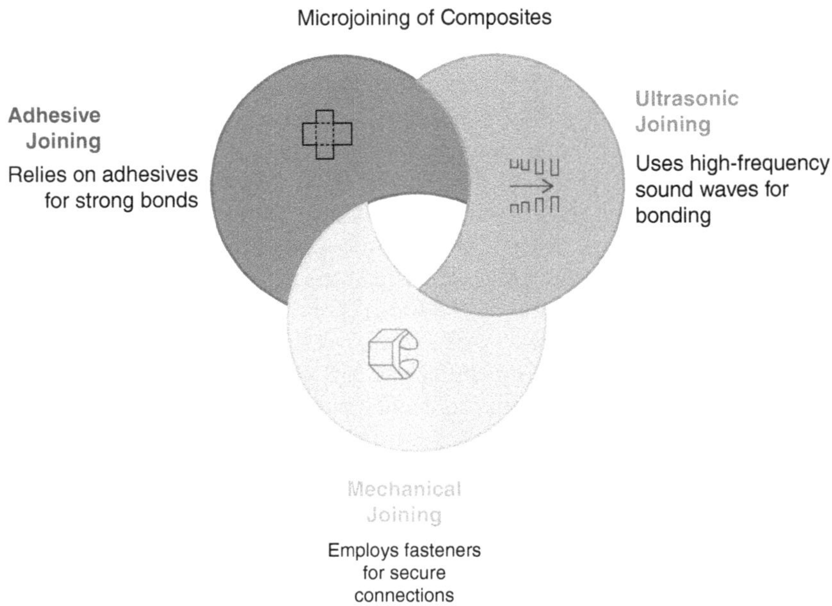

FIGURE 3.9 Composites-based micro-joining.

welding, vibrational energy typically accumulates near surface asperities, facilitating heat dissipation. Triangular protrusions known as "energy directors" or "susceptors" are molded onto the component to initiate melting. Tateishi[74] and Zach[75] proposed the use of tie layers as an alternative to energy directors in ultrasonic welding. A tie layer is an interlayer composed of substrate material that has been modified to lower the melting temperature and promote preferential melting.

3.7.2 Mechanical Joining

Mechanical joining, a crucial category of ceramic matrix composite joining technologies, is vital in numerous engineering applications. It often links CMCs to other components using fasteners such as bolts and rivets to create a cohesive structure. A common method to connect CMCs is through the use of bolts. It can accommodate substantial tensile, shear, and other loads, making it suitable for structural connections that require significant weight support.[76] Nevertheless, as CMCs are generally employed in environments characterized by elevated temperatures and pressures, bolted connections must adhere to more stringent criteria, which should be considered during their design and application. The difference in thermal expansion between CMCs and high-temperature alloys resulted in a significant reduction in preload and an adverse alteration in bolt-hole clearance. Temperature significantly influenced the propagation of damage and the tensile capacity of the joint construction. Zhao et al.[77] built a finite element framework and utilized gradual damage analysis to examine the impact of geometric parameters on failure loads and damage mechanisms at 750°C.

At 750°C, the connection structure had a low failure load due to the difference in plate edges and center-to-center spacing between the double bolts. Increasing the distance between the plate edges will result in a marginally stiffer connection and a more robust joint under identical failure forces. Varying the distance between the centers of two bolts resulted in distinct failure types of CMCs. The various failure modes interacted to alter the failure loads.

3.7.3 Rivet Joining Technology

The riveting process is commonly employed to join CMCs in specific domains. The tensile strength of rivets is limited due to the majority of force transmission between them and the hole walls being reliant on friction and mechanical interlocking. It is vulnerable to shear deformation or pull-out when subjected to substantial tensile or shear loads. Xu et al.[78] studied how C/SiC riveted joints fail under thermomechanical coupling loads. The brittle fracture of the interface layer between the rivets and the plates was found to be the main cause of failure in C/SiC riveted joints. The progressive degradation primarily manifested on the surface of the C/SiC composite material during the failure phase due to friction from the interference fit. The external temperature reaching 800°C had a negative effect on the out of plane ultimate strength of the C/SiC riveted joints. Xu et al.[78] investigated the impact of temperature fluctuations and the residual stress from riveting interference fits on the properties of C/SiC composite materials using both experimental and simulation analysis methods. The release of residual stress enhanced interface performance, resulting in a transition in the fracture cross-sectional shape from twisted fiber pull-out to a clean, brittle fracture. The rivets enhanced the load-bearing capacity of C/SiC composite materials marginally at ambient temperature. However, as the external temperature increased, the capacity to bear weight significantly diminished.

3.7.4 Adhesive Joining

Adhesive bonding is a crucial method for ceramic matrix composites (CMCs). It employs adhesives to exert forces that maintain the bonding of two or more surfaces, ensuring the efficacy of the connection.[79] This type of connection facilitates the assembly of components with diverse shapes and materials. Epoxy adhesive bonds aluminum alloys and CMCs. Upon completion of the procedures, comprising 12 hours at ambient temperature and 5 hours at 90°C, the shear strength of the joint was determined to be 31.6 ± 7.2 MPa. After 10 minutes of heating to 200°C, the joint's microscopic structure and shear strength increased by about 21%. The mechanical properties of the connected samples remained unchanged during thermal cycling. The samples were much weaker after the salt spray test, but the results at 240 and 720 hours did not vary statistically. Sandblasting enhances the roughness and durability of an aluminum alloy's surface. Anodized aluminum alloys exhibit improved corrosion and wear resistance; however, this process inhibits adhesive bonding, hence reducing shear strength, particularly when heat treatment takes place. Yan et al.[80] examined the effect of ultrasonic vibration on the bonding joints of CMCs. The suggested hydrodynamic model and capillary rise model may accurately predict the adhesive pressure and penetration capacity. Ultrasonic vibration increases the maximum hydrodynamic pressure of adhesives by five to seven orders of magnitude relative to the absence of vibration. This facilitates the adhesive's penetration into the microstructure of the bonding surface and enhances interfacial bonding. The mean shear strength of bonded joints subjected to ultrasonic vibration is 15.52 MPa. The mean shear strength of untreated joints is 12.96 MPa. The ultrasonic treatment enhances joint strength by 19.8% and improves their mechanical properties. Ionotropic gelation is an innovative method for bonding oxide CMCs. The efficacy of bonding Nextel 610/alumina–zirconia composites during different stages of manufacture was assessed. When the gel or green condition is applied, the joint functions at its finest. Both the composite plate and the connection have the same appearance and characteristics as the matrix of the composite plate. The shear strength of the junction is around 8 MPa, which is comparable to the composite plate's interlaminar shear strength. It is effective when combined with other substances at high temperatures. Additionally, no additional heat treatment is required for joints in the green stage. Shear strength is reduced as a result of voids and cracks at the joint contact following sintering. Utilizing the Refractory Eutectic Assisted Bonding (REABond) technique, we examined the joining of silicon carbide-based composites in the silicon-chromium, silicon-titanium, and silicon-hafnium systems when silicon-rich refractory eutectic phase compositions are employed.[81] This entailed employing silicon-hafnium (Si-Hf) eutectic phase material in tape form to create joints that were homogeneous, dense, well-adhered, and devoid of pores or voids. The joints fabricated using Si-Cr and Si-Ti eutectic pastes exhibited inconsistencies, characterized by unbonded regions, microcracks, and significant voids. The fiber orientation of the composites was correlated with the shear strength of the bonded specimens. When the fibers were aligned parallel to the joint position, the composites fractured. Fractures occurred when fibers were placed perpendicularly to the joint.[82]

3.8 SUMMARY

In summary, MJ refers to sophisticated techniques for assembling components that range from hundreds of nanometers to several hundred micrometers in size. This is crucial in domains such as electronics, aerospace, medical devices, and MEMS, where precision and compactness are essential. The methods include solid-state bonding (diffusion, friction stir, and ultrasonic), fusion welding (laser and electron beam welding), soldering/brazing, and adhesive bonding. Laser MJ is favored for its precision, versatility, and minimal thermal impact. CS and explosive welding enable the amalgamation of dissimilar or temperature-sensitive materials without inducing melting. Various materials require distinct methods for their assembly. Polymers employ extrusion, infrared welding, and ultrasonic welding; ceramics utilize brazing (active or microwave-assisted) and ultrasonic welding; glasses implement laser welding, active brazing, and ultrasonic-assisted brazing; and composites adopt adhesive bonding, refractory eutectic bonding, mechanical fastening, and ultrasonic welding. Each solution addresses issues related to joint strength, thermal discrepancies, and material compatibility in compact systems.

REFERENCES

[1] Pocius, A.V. 2002. *Adhesion and Adhesive Technology: An Introduction*, 2nd ed. Carl Hanser Gardener Verlag, Munchen.

[2] Hakala, K. and Tiainen, T. 2001. The general properties of polymeric Adhesives. In: *The Proceedings of Structural Adhesives in Engineering VI*, Bristol, UK, p. 143.

[3] Adams, R.D. 2005. *Adhesive Bonding: Science, Technology, and Applications*. Taylor & Francis.

[4] Haghshenas, M., Abdel-Gwad, A., Omran, A.M., Gökçe, B., Sahraeinejad, S. and Gerlich, A.P. 2014. Friction Stir Weld Assisted Diffusion Bonding of 5754 Aluminum Alloy to Coated High Strength Steels, *Materials & Design*, 55, pp. 442–449,

[5] Kah, P., Suoranta, R., Martikainen, J. and Magnus, C. 2014. Techniques for Joining Dissimilar Materials: Metals and Polymers. *Materials Science* 36, pp. 152–164.

[6] Martinsen, K., Hu, S.J. and Carlson, B.E. 2015. Joining of Dissimilar Materials. *CIRP Annals - Manufacturing Technology* 64, pp. 679–699.

[7] Settineri, L., Atzeni, E. and Ippolito, R. 2010. Self-Piercing Riveting for Metal-Polymer Joints. *International Journal of Material Forming* 3(Suppl 1), pp. 995–998.

[8] Hetrick, E.T., Baer, J.R., Zhu, W., Reatherford, L.V., Grima, A.J., Scholl, D.J., Wilkosz, D.E., Fatima, S. and Ward, S.M. 2009. Ultrasonic Metal Welding Process Robustness in Aluminum Automotive Body Construction Applications. *Welding Journal* 88, pp. 149–158.

[9] Balle, F., Wagner, G. and Eifler, D. 2007. Ultrasonic Spot Welding of Aluminum Sheet/Carbon Fiber Reinforced Polymer-Joints. *Materialwissenschaft und Werkstofftechnik* 38, pp. 934–938.

[10] Li, H., Sun, D., Cai, X., Dong, P. and Gu, X. 2013. Laser Welding of TiNi Shape Memory Alloy and Stainless-Steel Using Co Filler Metal. *Optics & Laser Technology* 45, pp. 453–460.

[11] Ageorges, C., Ye, L. and Hou, M. 2001. Advances in Fusion Bonding Techniques for Joining Thermoplastic Matrix Composites: A Review. *Composites Part A: Applied Science and Manufacturing* 32, pp. 839–857.

[12] Potente, H., Schnieders, J. and Bornemann, M. 2002. Theoretical Model for the One-Dimensional Temperature and Stress Calculation of Simple Hot Plate Welded Geometries. *Macromolecular Materials and Engineering* 287, p. 843.

[13] Schmachtenberg, E. and Tüchert, C. 2003. Long-Term Properties of Butt-Welded Poly(propylene). *Macromolecular Materials and Engineering* 288, p. 291.

[14] Kapoor, R.R., Podszus, E.S. and Eagar, T.W. 1988. Wettability of Silver-Based Reactive Metal Brazing Alloys on Alumina. *Scripta Metallurgica* 22(8), pp. 1277–1279.

[15] Niwa, Y., Kawahito, Y., Kubota, S. and Katayama, S. 2008. Evolution of LAMP Joining to Dissimilar Metal Welding. *Proceedings of ICALEO 2008, LIA*, Temecula, CA, pp. 311–317.

[16] Katayama, S. and Kawahito, Y. 2008. Laser Direct Joining of Metal and Plastic. *Scripta Materialia* 59(12), pp. 1247–1250.

[17] Abibe, A., Amancio-Filho, S., Dos Santos, J. and Hage, E. 2011. Development and Analysis of a New Joining Method for Polymer-Metal Hybrid Structures. *Journal of Thermoplastic Composite Materials* 24, pp. 233–249.

[18] McKnight, S.H., Holmes, S.T., Gillespie, J.W. Jr., Lambing, C.L.T. and Marinelli, J.M. 1997. Scaling Issues in Resistance Welded Thermoplastic Composite Joints. *Advances in Polymer Technology* 16, pp. 279–295.

[19] Ageorges, C., Ye, L. and Hou, M. 2000. Experimental Investigation of the Resistance Welding for Thermoplastic-Matrix Composites. Part I: Heating Element and Heat Transfer. *Composites Science and Technology* 60(7), pp. 1027–1039.

[20] Collins, E.A., Bareš, J. and Billmeyer, F.W. 1973. *Experiments in Polymer Science*. Wiley, New York.
[21] Cahn, R.W. and Haasen, P. 1996. *Physical Metallurgy*, 4th rev. and enhanced ed. North-Holland, Amsterdam.
[22] Tamjidy, B.T.M., Baharudin, H.T., Paslar, S., Matori, K.A., Sulaiman, S. and Fadaeifard, F. 2017. Multi-Objective Optimization of Friction Stir Welding Process Parameters of AA6061-T6 and AA7075-T6 Using a Biogeography Based Optimization Algorithm. *Materials* 10, p. 533.
[23] Amancio-Filho, S.T. and Dos Santos, J.F. 2009. Joining of Polymers and Polymer–Metal Hybrid Structures: Recent Developments and Trends. *Polymer Engineering & Science* 49(8), pp. 1461–1476.
[24] Yang, Y., Su, X., Li, J., Li, Y., Wang, S. and Liu, Y. 2017. Effects of Ti Content on Microstructure and Mechanical Properties of Ag–Cu–Ti Brazing Alloys. *Ceramics International* 43(4), pp. 3625–3631.
[25] Ageorges, C. and Ye, L. 2002. *Fusion Bonding of Polymer Composites*. Springer, London.
[26] Yarlagadda, P.K.D.V. and Tan, C.C. 1998. An Investigation into Welding of Engineering Thermoplastics Using Focused Microwave Energy. *Journal of Materials Processing Technology* 74, p. 199.
[27] Iyogi, K., Nakahashi, M., Takeda, H. and Shirokane, M. 1990. Brazing Paste for Bonding Metal and Ceramic. *US Patent 5019187 A*.
[28] Hao, H., Wang, Y., Jin, Z. and Wang, X. 1997. Interfacial Morphologies between Alumina and Silver-Copper-Titanium Alloy. *Journal of Materials Science* 32(18), pp. 5011–5015.
[29] Meek, T.T. and Blake, R.D. 1986. Ceramic—Ceramic Seals by Microwave Heating. *Journal of Materials Science Letters* 5(3), pp. 270–274.
[30] Nicholas, M.G., Valentine, T.M. and Waite, M.J. 1980. The Wetting of Alumina by Copper Alloyed with Titanium and Other Elements. *Journal of Materials Science* 15(9), pp. 2197–2206.
[31] Silberglitt, R., Ahmad, I., Black, W.M. and Katz, J.D. 1993. Recent Developments in Microwave Joining. *MRS Bulletin* 18(11), pp. 47–50.
[32] Cai, J., Xie, X.M., Baoshun, L., Huang, X.X., Chen, T.G. and Guo, J.K. 1992. Microwave Joining of $Bi_{1.6}Pb_{0.4}Sr_2Ca_2Cu_3O_x$ Superconductors. *Superconductor Science and Technology* 5(10), pp. 599–601.
[33] Chen, C., Wang, C., Chang, H. and Yeh, J. 2001. Microwave Joining of Alumina Ceramics Using Glass Frit. *Ceramics International* 27(4), pp. 345–349.
[34] Vandeputte, G., Oprins, J., Desplentere, F., Van Hemelrijck, D. and De Baets, P. 2012. Microwave Joining of Metal–Ceramic Joints Using a Reactive Interlayer. *Materials & Design* 36, pp. 664–670.
[35] Wang, J., Liu, X., Ma, X. and Wang, S. 2013. Effect of Titanium Content on the Microstructure and Mechanical Properties of Ag–Cu–Ti Brazing Alloys. *Journal of Alloys and Compounds* 579, pp. 90–96.
[36] Zhang, X., Zhang, H., Huang, Z. and Yang, X. 2010. Brazing of Alumina Ceramics to Titanium Alloy Using Ag–Cu–Ti Filler Metal. *Journal of Materials Science & Technology* 26(11), pp. 1055–1060.
[37] Lee, S.K., Kim, H.S. and Kim, Y.W. 2009. Wetting and Interfacial Reactions of Ti-Containing Brazing Alloys on Alumina Substrates. *Materials Chemistry and Physics* 113(2), pp. 559–565.
[38] Yu, L., Wang, Y., Sun, Y., Chen, W. and Xu, Y. 2017. Influence of Brazing Parameters on Microstructure and Mechanical Properties of Ag–Cu–Ti/Al_2O_3 Joints. *Materials Science and Engineering: A* 696, pp. 72–80.

[39] Wang, Y., Hao, H., Liu, H. and Jin, Z. 1999. Interfacial Reactions of Ag–Cu–Ti Alloy with Alumina. *Journal of Materials Science Letters* 18(11), pp. 841–843.
[40] Choi, S.W., Kang, J.K., Park, H.J. and Kim, J.G. 2011. Microstructure and Mechanical Properties of Ag–Cu–Ti Brazed Alumina Ceramics. *Materials Science and Engineering: A* 528(3), pp. 1445–1450.
[41] Su, X., Liu, Y., Li, Y., Li, J., Yang, Y. and Wang, S. 2015. Effects of Ti Content on Microstructure and Mechanical Properties of Ag–Cu–Ti Brazing Alloys for Alumina Ceramics. *Ceramics International* 41(7), pp. 8660–8667.
[42] Chen, Q., Gao, L., Yang, C., Xie, Z. and Zhang, Y. 2014. Brazing of Alumina Ceramics to Stainless Steel Using Ag–Cu–Ti Alloy. *Journal of Materials Science & Technology* 30(9), pp. 909–914.
[43] Wang, S., Yang, L., Zhang, H., Guo, X. and Liu, X. 2018. Microstructure and Mechanical Properties of Ag–Cu–Ti Brazed Alumina Ceramics. *Materials Science and Engineering: A* 712, pp. 278–285.
[44] Xu, Y., Yu, L., Chen, W. and Wang, Y. 2019. Influence of Ti Content on Wetting Behavior and Microstructure of Ag–Cu–Ti Brazing Alloy on Alumina Ceramic. *Journal of Alloys and Compounds* 797, pp. 60–67.
[45] Huang, Z., Li, F., Zhang, X. and Yang, X. 2012. Brazing of Alumina to Titanium Alloy Using Ag–Cu–Ti Filler Metal: Microstructure and Mechanical Properties. *Materials & Design* 35, pp. 265–270.
[46] Lin, K., Singh, M. and Asthana, R. 2014. Interfacial Characterization of Alumina-to-Alumina Joints Fabricated Using Silver-Copper-Titanium Interlayers. *Materials Characterization* 90, pp. 40–51.
[47] Chen, W., Wang, Y., Yu, L., Sun, Y. and Xu, Y. 2018. Influence of Brazing Time on Microstructure and Mechanical Properties of Ag–Cu–Ti Brazed Alumina Ceramics. *Materials Science and Engineering: A* 717, pp. 255–261.
[48] Li, J., Yang, Y., Su, X., Wang, S., Liu, Y. and Li, Y. 2020. Microstructure and Mechanical Properties of Ag–Cu–Ti Brazed Alumina Ceramics as a Function of Ti Content. *Ceramics International* 46(2), pp. 1860–1866.
[49] Park, J.S., Lee, H.S., Kim, M.S. and Choi, W.K. 2019. Influence of Brazing Temperature and Time on Microstructure and Shear Strength of Ag–Cu–Ti Brazed Alumina. *Journal of Materials Processing Technology* 270, pp. 120–127.
[50] Liu, X., Zhang, Y., Wang, Z. and Zhao, H. 2017. Effect of Ti Content on Wetting and Mechanical Properties of Ag–Cu–Ti Brazing Alloy for Alumina Ceramics. *Ceramics International* 43(8), pp. 6008–6015.
[51] Li, X., Zhang, W., Zhou, X. and Liu, J. 2016. Effect of Brazing Parameters on Microstructure and Mechanical Properties of Ag–Cu–Ti Brazed Alumina Ceramics. *Ceramics International* 42(9), pp. 10805–10811.
[52] Kim, S.H., Lee, J.H. and Park, J.S. 2014. The Effect of Brazing Temperature on Microstructure and Mechanical Properties of Ag–Cu–Ti Brazed Joints Between Alumina and Stainless Steel. *Ceramics International* 40(8), pp. 12365–12371.
[53] Choi, Y.J., Lim, H., Kim, J.G. and Lee, S.K. 2013. Wetting and Bonding Behavior of Ag–Cu–Ti Alloy on Alumina Substrate. *Materials Chemistry and Physics* 139(1), pp. 431–436.
[54] Zhang, Y., Li, H., Chen, Q., Yang, C. and Xie, Z. 2015. Effects of Brazing Atmosphere on Microstructure and Mechanical Properties of Ag–Cu–Ti Alloy Brazed Alumina Ceramics. *Journal of Materials Science & Technology* 31(3), pp. 307–313.
[55] Yang, L., Wang, H., Liu, X., Guo, X. and Zhang, H. 2018. Study on the Interfacial Reaction and Mechanical Properties of Ag–Cu–Ti Brazed Alumina Ceramics. *Materials Science and Engineering: A* 712, pp. 234–240.

[56] Huang, Z., Tang, M., Wu, X. and Zhao, J. 2020. Investigation of Microstructure and Shear Strength of Ag–Cu–Ti Brazed Alumina Ceramics Joints. *Ceramics International* 46(11), 17234–17241.

[57] Park, S., Kim, D. and Lee, H. 2018. Optimization of Brazing Parameters for High Strength Ag–Cu–Ti Brazed Alumina Ceramic Joints. *Ceramics International* 44(12), pp. 14678–14685.

[58] Chen, Z., Liu, Y., Yang, W., Zhang, H. and Li, S. 2021. Microstructure Evolution and Mechanical Properties of Ag–Cu–Ti Brazed Alumina Ceramic Joints. *Materials* 14(7), Article 1673.

[59] Sun, Y., Wang, X. and Zhao, J. 2018. Wettability and Bonding Strength of Ag–Cu–Ti Brazing Alloy on Alumina Ceramics under Different Brazing Temperatures. *Ceramics International* 44(1), pp. 245–252.

[60] Wang, Y., Zhao, Y., Xu, F., Liu, J. and Zhang, C. 2018. Effect of Ti Content on the Wettability and Microstructure of Ag–Cu–Ti Alloy Brazed Alumina Ceramics. *Journal of the European Ceramic Society* 38(15), pp. 5247–5254.

[61] Kim, D., Park, S. and Lee, K. 2020. Optimization of Brazing Parameters for Ag–Cu–Ti Alloy and Alumina Ceramic Joint. *Ceramics International* 46(1), pp. 1003–1010.

[62] Xu, Y., Sun, Y., Chen, W., Yu, L. and Wang, Y. 2019. Microstructure and Mechanical Properties of Ag–Cu–Ti Brazed Alumina Ceramic Joints as a Function of Brazing Atmosphere. *Journal of Materials Science & Technology* 35(4), pp. 631–638.

[63] Zhang, C., Li, J., Yang, Y., Su, X. and Wang, S. 2017. Effect of Brazing Atmosphere on Wettability and Microstructure of Ag–Cu–Ti Brazing Alloy on Alumina Ceramic. *Ceramics International* 43(15), pp. 11630–11637.

[64] Li, H., Chen, Q., Zhang, Y., Yang, C. and Xie, Z. 2015. Effects of Brazing Atmosphere and Ti Content on Microstructure and Mechanical Properties of Ag–Cu–Ti Brazed Alumina Ceramics. *Journal of Alloys and Compounds* 650, pp. 314–321.

[65] Choi, Y.J., Lim, H., Kim, J.G. and Lee, S.K. 2014. Effect of Brazing Parameters on Microstructure and Mechanical Properties of Ag–Cu–Ti Alloy Brazed Alumina Ceramic Joints. *Materials Chemistry and Physics* 144(3), pp. 995–1001.

[66] Sun, Q., Li, X., Wang, D. and Feng, J. 2019. Effect of Brazing Time on the Microstructure and Mechanical Properties of Ag–Cu–Ti Brazed Alumina Joints. *Journal of the European Ceramic Society* 39(4), pp. 1305–1312.

[67] Zhao, W., Yu, Z. and Ma, J. 2017. Wettability and Interfacial Reactions of Ag–Cu–Ti Alloy on Alumina Substrate. *Materials Letters* 197, pp. 76–79.

[68] Lee, S.H., Park, J. and Kim, D. 2021. Influence of Brazing Atmosphere on the Microstructure and Bonding Strength of Ag–Cu–Ti/Alumina Joints. *Journal of Materials Science & Technology* 64, pp. 102–110.

[69] Wang, K., Chen, F. and Li, Y. 2018. Microstructural Evolution and Mechanical Properties of Ag–Cu–Ti Brazed Alumina Joints under Different Brazing Conditions. *Ceramics International* 44(15), pp. 18018–18025.

[70] Yang, J., Liu, X. and Zhao, Y. 2019. Effect of Ti Content on Wetting Behavior and Mechanical Performance of Ag–Cu–Ti Brazing Alloy for Alumina Ceramics. *Materials Science and Engineering: A* 752, pp. 214–220.

[71] Kim, H., Lim, J. and Park, S. 2020. Brazing Behavior and Strength of Ag–Cu–Ti/Alumina Joints under Various Temperature Conditions. *Ceramics International* 46(2), pp. 1772–1779.

[72] Liu, J., Zhao, L. and Zhang, Q. 2016. Influence of Brazing Atmosphere on the Formation of Interfacial Reaction Layer in Ag–Cu–Ti Brazed Alumina Ceramic Joints. *Journal of Alloys and Compounds* 689, pp. 85–90.

[73] Chen, W., Sun, Y. and Wang, Y. 2017. Microstructure and Mechanical Properties of Ag–Cu–Ti Brazed Alumina Ceramic Joints Prepared under Vacuum. *Materials* 10(10), p. 1145.
[74] Zhao, X., Wang, J. and Liu, H. 2021. Interfacial Characterization and Mechanical Properties of Ag–Cu–Ti Brazed Alumina Ceramics under Different Brazing Pressures. *Materials Science and Engineering: A* 799, p. 140271.
[75] Gao, F., Chen, Y. and Zhang, M. 2019. Effects of Post-Brazing Heat Treatment on the Microstructure and Shear Strength of Ag–Cu–Ti/Alumina Joints. *Ceramics International* 45(5), pp. 6420–6427.
[76] Liu, Y., Xu, Q. and Jiang, Z. 2017. Influence of Brazing Gap on Microstructure and Mechanical Properties of Ag–Cu–Ti Brazed Alumina Ceramic Joints. *Journal of Materials Processing Technology* 244, pp. 138–146.
[77] Park, J., Lee, S. and Kim, H. 2020. Microstructural Analysis of Ag–Cu–Ti Brazed Alumina Ceramics Joints Using SEM and EDS. *Materials Characterization* 167, p. 110487.
[78] Wang, L., Zhao, P. and Chen, F. 2019. Effect of Ti Content in Ag–Cu–Ti Brazing Alloy on the Formation of Interfacial Reaction Layer with Alumina. *Ceramics International* 45(12), pp. 15635–15641.
[79] Zhang, H., Liu, S. and Tang, Y. 2016. Influence of Brazing Atmosphere on Microstructure and Mechanical Properties of Ag–Cu–Ti/Alumina Joints. *Ceramics International* 42(14), pp. 15840–15847.
[80] Chen, X., Li, Y. and Wang, D. 2017. Investigation of Microstructure and Shear Strength of Ag–Cu–Ti Brazed Alumina Ceramic Joints under Vacuum. *Journal of the European Ceramic Society* 37(2), pp. 489–495.
[81] Liu, Q., Zhang, X. and Sun, J. 2018. Brazing Temperature and Holding Time Effects on the Microstructure and Mechanical Properties of Ag–Cu–Ti/Alumina Joints. *Ceramics International* 44(18), pp. 22486–22494.
[82] Lim, S., Kim, D. and Park, J. 2020. Optimization of Brazing Parameters for Ag–Cu–Ti Alloy Joining Alumina Ceramic with Copper Substrate. *Materials & Design* 190, p. 108577.

4 Laser Micro-Welding for Electronics and Semiconductor Applications

4.1 INTRODUCTION

Laser micro-welding is a precision welding process that utilizes laser beams to weld minute parts or materials. To this day, reliability is the major driving factor for adoption of laser micro-welding. A laser is the ideal tool for welding with minimal spot sizes, very high-energy density, such as excellent control over energy management and ease of access.[1] Lasers enable the fastening of electronic devices together, which ensures very reliable production of electrical components and semiconductor applications. Key laser micro-welding application aspects include a range of the following:

(a) **Precision Welding:**
A common example of laser micro-welding is its ability to provide ultra-precise and concentrated heat input. This is quite important when dealing with small electronic components in semiconductor devices.

(b) **Minimal Heat-Affected Zone (HAZ):**
Laser micro-welding is used in the industrial process to create low HAZ, thus making it possible to work on very fine electric components. This maintains the integrity of the semiconductor devices and reduces the probability of nearby components being damaged.

(c) **Material Compatibility:**
Laser micro-welding is also employed in semiconductors and electronics, joining together materials such as metals, alloys, and certain types of polymers. Such flexibility helps when bonding dissimilar materials so common in semiconductor manufacturing.

(d) **High Speed and Efficiency:**
It is usually a fast and efficient process, allowing for high-throughput production by semiconductor manufacturers. This is necessary in mass manufacturing even in the electronics sector.

(e) **Automation Integration:**
Automated manufacturing lines may readily include laser micro-welding. This integration guarantees constant weld quality and improves the production process' overall efficiency.

DOI: 10.1201/9781003502395-4

(f) **Fine Feature Welding:**
Small-scale components and tiny features can be welded together because of the ability of laser beams to focus on extremely small spot sizes. This is critical to the electronics industry's tendency toward miniaturization.

(g) **Hermetic Sealing:**
In semiconductor packaging, laser micro-welding is frequently used for hermetic sealing. The interior components are guaranteed to be shielded from external elements like dampness and pollutants thanks to this sealing.

(h) **Reduced Contamination:**
Compared to conventional welding techniques, laser welding has a lower chance of contamination because it is a noncontact operation. This is especially crucial in the semiconductor production process, where cleanliness is paramount.

(i) **Customization and Flexibility:**
High levels of customization are possible with laser micro-welding systems, enabling producers to modify the procedure to meet particular needs. This adaptability is useful in the ever developing electronics industry.

(j) **Quality Control:**
Quality control procedures may be implemented more easily using laser micro-welding. It is simpler to monitor and guarantee the quality of the welds in electronic components due to the process's accuracy and reproducibility.

To sum up, laser micro-welding is a flexible and effective method that satisfies the stringent needs of precision welding in semiconductor and electronics applications. Its capacity to produce heat that is high-quality, regulated, and localized makes it suitable for contemporary electronic gadgets.

4.2 MICRO-ELECTRONICS WIRE BONDING

Wire bonding is one of the techniques for making electrical connections between semiconductors and silicon chips via bond wires. The bond wires are only thin gold or aluminum wires. The two most used techniques are aluminum wedge bonding and gold ball bonding.

The wire bonding process can be described as electrical joining of hundreds or thousands of tiny wires using very small pressure, heat and ultrasonic energy. Essentially, two basic materials are used during the process a wire and pad that are kept close to each other than subsequently fuse in a way similar to welding as examples. In other words, it results in the making of wire bonds along with flowing of electrons from one end to another. An example of a schematic fabrication process including bonding of wire is represented in Figure 4.1.

4.2.1 WIRE BONDING MECHANISMS

Three distinct conventional wire bonding mechanism approaches are available according on the kind of input. Thermocompression bonding, ultrasonic bonding, and thermosonic bonding are the significant mechanisms.

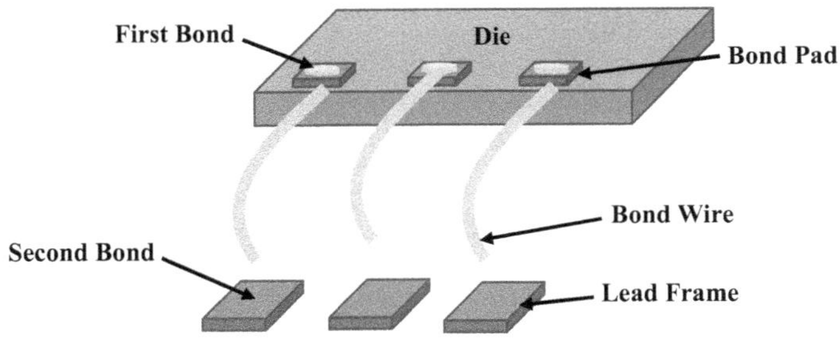

FIGURE 4.1 Mechanism of bonding wire.[1]

4.2.1.1 Thermocompression Bonding

A mechanical force with high temperature is used in the conventional process of thermocompression bonding. The gold (AU) wire is commonly used for wedge-wedge bonding to gold pads. The two pieces of metal, i.e., the bonded wire and substrate metallization or pad, come into close proximity over a predetermined period of time, temperature, and pressure, and thermocompression bonding takes place. During the cycle of bonding, the materials are plastically deformed, and they undergo atomic interdiffusion to some degree. A homogeneous gold interface may be produced by this atomic interdiffusion if substrate metallization, gold pad, and gold wire are utilized together.[2] These days, because of its lengthy processing durations and high-temperature requirements, it is not recommended.

4.2.1.2 Ultrasonic Bonding

Ultrasonic bonding employs both ultrasonic energy and mechanical force to carry out the procedure. This kind is used for wedge-wedge bonding in high-power electronic applications to connect the aluminum wire to aluminum or gold pads. Wedge-wedge is a bonding procedure that occurs at low temperatures. In this process, a transducer is used to vibrate the bonding tool (wedge) at frequencies ranging from 20 to 300 kHz, providing the energy required for metal welding. The predominant frequency remains 60 kHz. During ultrasonic bonding, the wedge's tip oscillates in a direction running in the same direction as the bonding pad. Ultrasonic bonds are commonly created using aluminum or aluminum alloy wire, which is applied to either aluminum or gold pads. Ultrasonic bonding with gold wire has been conducted using both ribbon and round wire, however its usage is not widespread. Several metals like copper and other have been joined using the ultrasonic bonding procedure, particularly in certain applications. Ultrasonic wedge bonding has several significant benefits. It may create a robust connection without requiring much or any heat to be delivered to the substrate. Additionally, it can usually be done at smaller intervals than ball bonding techniques.[2]

4.2.1.3 Thermosonic Bonding

The process involves combining thermocompression and ultrasonic bonding techniques to create a gold wire bonded using ball-stitch technique. Thermosonic bonding is the favored technique for connecting integrated circuit chips in packaging, as it minimizes mechanical strain and processing duration.

TABLE 4.1
Process Variables in Wire Bonding Mechanism

Type of Wire Bonding Process Used	Pressure	Temperature	Ultrasonic Energy	Type of Wire Material Used
Thermocompression Bonding	Higher	300°C–500°C	Doesn't Require	Gold
Ultrasonic Bonding	Lower	25°C	Required	Gold, Aluminum
Thermosonic Bonding	Lower	100°C–150°C	Required	Gold

Thermosonic bonding uses of ultrasonic energy and ball bonding capillary technique of thermocompression bonding. The thermosonic process is often conducted similar to the thermocompression process; however, the capillary is not heated. Substrates are usually maintained at temperatures ranging from 100°C to 150°C. When the wire and the capillary touch the pad, brief pulses of energy (lasting tens of milliseconds) are used on the capillary. Interconnection between the substrates and ICs takes place at a temperature below 150°C. This allows them to be securely attached using epoxy/organic adhesives, avoiding degradation caused by continued exposure to temperature that are higher than the temperatures at which materials undergo a transition from a rigid, glassy state to a more flexible one, which could occur due to excessively high temperatures. Moreover, the likelihood of uncontrolled intermetallic development is greatly reduced. Thermosonic bonding is mainly performed with gold wire; however, effective demonstrations have also been made with aluminum, copper, and palladium thermosonic bonding.

Thermosonic gold ball bonding is now the dominant technology due to its superior speed compared to ultrasonic aluminum bonding. After the ball bond, the wire is freely moved in any direction, causing no strain on the wire. This substantially simplifies the automation of wire bonding since the movement just has to be in the x and y axes. Table 4.1 shows the process variables related to the three wire bonding procedures.

4.2.2 Wire Bonding Materials Used

The material that is used for bonding must exhibit excellent electrical conductivity, adhesion, resistance to corrosion and hardness. The following is a compilation of several materials and their respective combinations:

(a) **Aluminum:**
It is appropriate for thermosonic bonding. Alloyed aluminum wires are favored because of their ultimate dimensions and increased durability. Some commercially utilized combinations of wire-pad materials include Al-Au, Al-Ni, and so on. Aluminum in its pure form is not sufficiently hard to be drawn; hence, it is combined with either 1% Si or 1% Mg to enhance its strength.

(b) **Copper:**
It is a highly regarded material for use in applications of semiconductors. Wires made of copper possess a substantial diameter and can serve as a substitute for aluminum wire in situations involving strong electrical currents. This technique is employed for precise wire ball bonding within the size range of 10 micrometers to 100 micrometers. The ball bonding process should be conducted in an inert environment due to the high reactivity of copper oxides.

(c) **Silver:**
Wires made of silver (Ag) exhibit excellent electrical and thermal conductivity. The appropriate wire-pad combinations are Ag-Sic, Ag-Au, and Ag-Al.

(d) **Gold:**
Ball bonding requires the use of pure gold wires that include enough quantities of other elements. Commonly employed combinations include Au-Al, Au-Cu, Au-Pd, and so on. Thermocompression and thermosonic bonding utilize gold wire. Optimal surface polish and surface cleanliness are crucial for promoting robust bond formation and reducing the obstruction of capillaries.

4.2.3 Types of Wire Bonding

Wire bonding is categorized into ball bonding and wedge bonding. Table 4.2 displays the process of wire bond creation, including the specific bonding technique, the kind of bonding tool utilized, the wire material type, the pad material type, and the speed at which the wire moves.

4.2.3.1 Ball Bonding

It is the predominant method for creating electrical connections.

The thermosonic ball bonding procedure of a gold wire to gold or aluminum pads is listed in the following steps:

TABLE 4.2
Wire Bond Formation[3]

Type of Bond	Bonding Technique	Bonding Tool	Type of Wire Material	Type of Pad Material	Speed
Ball Bond	Thermocompression/ Thermosonic	Capillary	Gold	Aluminum, Gold	10 wires/sec (Thermosonic)
Wedge Bond	Thermosonic/ Ultrasonic	Wedge	Gold, Aluminum	Aluminum, Gold	4 wires/sec
Ball Bond	Thermocompression/ Thermosonic	Capillary	Gold	Aluminum, Gold	10 wires/sec (Thermosonic)

1. A gold wire is threaded via a capillary with a ceramic bond, which is a needle-shaped instrument. At this point, a high voltage charge is applied to the wire, resulting in the melting of the wire and forming a gold sphere (free air ball or FAB).
2. Hereafter, the capillary tip will form a FAB where after the tool now moves to a designated bond pad on the heated device surface horizontally traced back right over it.
3. The device simply uses a prescribed amount of pressure to push the air ball against the pad Ultrasonic energy that is used to bond the pad and ball together.
4. It then moves to the next bond pad and does the stitch bond for that.
5. The wire is subjected to compression as it is squeezed between one side of the pad and the capillary tube.
6. These are the ultrasonic vibrations, force applied, and thermal energy that make the bond between pad and wire.
7. Grab the clamp of the wire securely and use the tool to pull this upwards in a vertical motion, thereby removing the wire together with it. A tiny tail from the wire sticks out of one end of the capillary, and each cycle starts by applying a high voltage electric charge to this tail.

4.2.3.2 Gold Ball Bumping

Gold ball bumping is a technique that is comparable to ball bonding and is employed to create gold-to-gold linkages for flip-flop packaging. The gold bumps are a wafer-level process (arcades, implant), which means they are stamped onto the ultra-thin layer of semiconductor material that is used to make integrated circuits (ICs).

1. To burn a hole in the wire, use an electrical flame-off method for melting the gold wire, and form it into a bowl without any ambient medium.
2. The FAB sticks to the pad using ultrasound applied with temperature and pressure.
3. Instead of running stitch bond, the tool rises to a particular altitude.
4. Then the wire is broken free and the tool is pulled upward.
5. Traditionally, the wafer is cut into smaller pieces, flipped over and bonded with gold pads by thermosonic bonding.

The ball bumping allows the bumps to be stacked vertically. The near-term application is the leveling of the upper bumps in what may be referred to as ball bumping. The process has been coined appropriately enough, and this method guarantees the same heights are achieved every time for both bumps and flattened surfaces, thus creating a larger area to bond.

4.2.3.3 Wedge Bonding

Ultrasonic bonding is a process that involves using aluminum wire to bind to aluminum or gold bond pads. Copper wedge bonding is capable of creating robust connections even at ambient temperature. The procedure for wedge bonding consists of the following steps:

1. The wire is passed through the wedge by use of the tool.
2. Next, the wire is applied to the bond pad with a specific amount of pressure. An ultrasonic transducer is used to vibrate the wedge in a parallel direction onto the substrate and parallel to the wire axis, utilizing ultrasonic energy. The frequency is within the range of 120–140 kHz.
3. The tool proceeds to the second bond position in the next phase. Resembling the first connection. The second bond is achieved with the utilization of ultrasonic energy and force.
4. By exerting a force in the opposite direction, the wire clamp is securely shut, and the wire is forcefully detached.

4.2.3.4 Aluminum Wedge Bonding

Aluminum is exclusively employed as a wire-pad material in aluminum wedge bonding. In this scenario, an aluminum wire that is securely fastened is brought into touch with the aluminum bond pad. The wire is secured to the bond pad by applying ultrasonic energy for a certain duration, resulting in the formation of the initial wedge bond. The second connection is created by additional use of ultrasonic energy on the wire. Subsequently, the wire is severed due to clamping. The use of ultrasonic wedge bonding with aluminum wire allows for the use of slender wire in fine-pitch applications and thick wire in applications requiring high power.

4.2.4 Operational Mechanism of a Wire Bonder

A wire bonder is a device that executes operations in accordance with the specifications of the process. Various wire bonders are offered in accordance with the bonding type, operational modes, and operational procedures. As needed, these devices may be operated manually, automatically, or semiautomatically.

4.2.4.1 Wire Bonding Machine

In addition to wedge bonding, ball bonding is also performed with a manual wire bonding machine. It increases operational stability, decreases bond pad pitch, and improves bonding rates. A wide variety of testing apparatus is utilized to monitor the efficacy of wire bonders. Through the utilization of the equipment's vision and positioning systems, the bonding location can be specified. As the bond forms, automated equipment measures the impedance of the bonding instrument in ambient air. These measurements have the potential to enhance the ultrasonic system's efficacy and assist in its troubleshooting. An additional attribute is the ability to seamlessly transition between products with minimal modifications to tools and software. The user has the ability to create software programs for a variety of objects, which can then be stored on a disk and utilized during the process.

4.2.4.2 Ribbon Bonding-Explanation

Ribbon bonding is an alternative method of interconnecting semiconductors to round wires by utilizing ribbons. Its use in high-frequency electronic applications is widespread. There are some justifications for employing ribbon bonding.

A charged free electron scatters other electrons uniformly as it traverses the conductor. Nevertheless, the wire is enveloped by a magnetic field whose strength is contingent upon the current flow direction. Self-induction occurs when the magnetic field of a conducting material induces an increase in separation pressure at its outer surface in response to a high-velocity signal. Primarily, current travels between the skin's outermost layer and a depth known as the skin depth. Depending on the frequency of the alternating current, the epidermis depth can differ. As the frequency rises, alternating current migrates toward the surface, resulting in a reduction in epidermis depth. The result is the rise in the effective resistance and a reduction in effective cross-sectional area.

A large surface area is provided by the ribbon relative to its cross-sectional area. As a consequence, increased power efficacy ensues. Ribbons enable the stacking of bonds that are advantageous for high-frequency applications. They provide enhanced mechanical loop stability. Wedge bonders are capable of manipulating both round and ribbon wires made of gold or aluminum. Technological advancements in ribbon bonding result in increased product efficiency and yield. Important factors for successful and dependable bonding include the material selection, size and shape of the bond, and the configuration of the wire coil. Presently, fold bonding is the method of choice throughout the semiconductor industry.

4.3 SOLID-STATE DIFFUSION BONDING FOR SEMICONDUCTORS

Diffusion bonding is an assembly technique employed in solid states to connect dissimilar and similar materials while ensuring minimal dimensional tolerances and granular microscopic deformation. Metalworking employs diffusion bonding that can join dissimilar and similar metals. During the solid-state diffusion (SSD), the atoms between the two metal surfaces solidify and move gradually. This is commonly accomplished at a temperature around 75% higher than the materials' absolute melting point. Bonding time, pressure, and temperature are the three critical variables that control the process of diffusion bonding. SSD bonding can be of the utmost importance when it comes to the development of dependable and high-performing semiconductor devices.

4.3.1 How Solid-State Diffusion Bonding Works

(a) **Pretreatment of Surfaces:**
The surfaces of the materials to be bonded are carefully prepared. Decontamination and roughening are commonly employed to achieve the needed levels of cleanliness and bondability.

(b) **Contact and Pressure:**
The two surfaces prepared are brought close, thereby making them contact. As a result of this pressure, the atoms or molecules at the interface in a material are forced to come closer.

(c) **Temperature:**
The temperature at which the materials are heated is usually below melting but high enough to cause diffusion of atoms. This enables atoms or molecules in the cooler atmosphere to move from one substance to another at contact.

(d) **Diffusion:**
The atoms or molecules of one substance start diffusing into the other material to form atomic bonds in the junction at higher temperatures. Pressure can assist the diffusion process by flattening down barriers at large surface imperfections, thus liberating diffusion pathways by increasing atomic mobility.

(e) **Solidification and Cooling:**
In the last step, after the sufficient bonding time is passed, constructed components are allowed to cool so that atoms or molecules on each sides form strong connections at borders and become solid from a different state.

(f) **Optional Post-Bonding Treatment:**
Depending on the application and materials, post-bonding treatments such as annealing can be performed to relieve stress and enhance bonding strength.

4.3.2 Applications in Semiconductor Manufacturing

(a) **Wafer Bonding:**
Wafer bonding is a standard technique used in semiconductor manufacturing with common use of SSD bonding. The process details the bonding of two semiconductor wafers to create complex structures or add different materials—e.g., in the manufacture of silicon-on-insulator (SOI) substrates or the making of MEMS (microelectromechanical systems) devices.

(b) **Heterogeneous Integration:**
SSD bonding enables the incorporation of different materials, e.g., silicon with III-V compound semiconductors as indium phosphide (InP) or gallium arsenide (GaAs). Thereby, fostering the progress of complex optoelectronic and photonic devices.

(c) **Interconnects:**
Solid-state stain diffusion bonding is one such method that may be used more commonly to create reliable interconnections between semiconductor devices or alternatively between semiconductor and nonsemiconductor materials, for instance. These interconnects are crucial to the fabrication of ICs and electronic packaging.

(d) **Heat Management:**
SSD bonding can be used to enhance the heat removal in high-power electronic devices, i.e., power amplifiers and RF transistors by attaching thermal conductive materials substrates with semiconductors species.

(e) **MEMS Fabrication:**
The field of Microelectromechanical systems (MEMs) uses SSD bonding to combine different materials in 3D structures with precise alignment and excellent mechanical stability.

SSD bonding plays an indispensable role in the fabrication process of advanced devices to ensure performance, reliability, and functions.

4.3.3 Overview

3D integration is a likely universal method to overcome the limits of Moore's law by extending the integration in the vertical direction. The use of a short link in 3D integration will dramatically reduce both the bandwidth and power dissipation, making the system more compact. 3D integration is made possible by 3D interconnect and can be achieved with a variety of bonding methods, such as direct bonding (DB), surface activated bonding (SAB), or eutectic bonding. Eutectic bonding is often used for wafer-to-wafer or chip-to-chip stacking due to its high process compatibility, meaning that it can tolerate deviations in surface topography and in the presence of particles, and does not necessitate ultrahigh vacuum conditions. This is especially the case when Pb-free Cu-Sn bonding is employed.

Initially, 3D connection technology bumps are obtained with pitch larger than 100 μm, and a small bump is created utilizing a conventional solder (like SnAg, SnAgCu, etc.) deposition technique. The conventional method for such bonding is a copper-tin bond using solder that allows 3D connection over broad pitch. In order to increase the density of reconfigured 3D integration in the future, bump pitch reduced to less than approximately 100 μm. The bump pitches are downscaled down up to similar ranges such as 20 μm or even smaller than 10 μm that are manufactured either with an electroplating technique or using physical vapor deposition (PVD) process. However, they are much less efficient and more expensive than electroplating at the same time. The bonding of Cu-Sn eutectic in a fine-pitch 3D interconnect can be conventionally performed either by the high-temperature, solid-liquid-interdiffusion (SLID) bonding, or by the low-temperature SSD bonding. SLID bonding with Cu-Sn requires a reducing process temperature of 250°C to 280°C and the temperature is slightly beyond the melting point of Sn (232°C). The Cu-Sn SLID bonding process experience an overflow of tin (Sn) quite easily, which can cause electrical short circuits between adjacent bumps. Therefore, the Cu-Sn SLID bonding can also be applied to 3D interconnects having bump pitch of several tens micrometers. The use of Cu-Sn SLID bonding leads to the same problems; thus, there is a need for an alternative method. The suggested solution is Cu-Sn SSD bonding, which ensures bump pitch being less than that provided by Cu-Cu direct contact bond. Cu-Sn SSD bonding is ideal for 3D interconnects with desired bump pitch of 10 to 30 μm due to the incapability of electroless plating to realize smaller bump pitch, while Cu bumps using pillar array have a planer top surface easily oxidized by forming thin oxide films. This oxide layer acts as a barrier, which prohibits the Cu-Sn from penetrating. In the Cu-Sn bonding of SSDs, few studies have performed surface pretreatments to remove the oxide layer and facilitate a reliable bond at low temperature. Cu-Sn bonding at the chip level has been studied at temperatures in the range of 150°C to 200°C. Pretreatment was done with a cleaning solution, dilute sulfuric acid until the Cu/Sn/Cu connection became conductive. The Cu/Sn microcone interface exists, where a SSD has been identified when the pitch is lowered less than 10 μm, employing PVD for bumping creation. An example of the ultrafinepitch 3D connection is Cu-Cu DB.

Fabrication and demonstration of a wafer-level sandwich interconnect with copper (2 μm)/tin (5 μm) layers are to be noted. Interconnect with a 20 μm bump pitch was successfully fabricated by using SSD bonding at a temperature of 200°C after removing the oxide layer and activating the bump surface before treatment. The bonding performance was characterized using techniques including structure and composition analysis at the interface, shear bond strength measurement, electrical testing and durability studies under thermal cycling (TC) or electromigration (EM).

4.3.4 Experiments

The process of achieving wafer-level, fine-pitch Cu/Sn/Cu interconnect involves four steps: test vehicle design, manufacture of fine-pitch bumps, pretreatment of the bump surface, and bonding of Cu/Sn/Cu SSD.

4.3.4.1 Test Vehicle Design

Bonding pairs are comprised of two wafers measuring 4 inches in diameter. 40 dies are present in each. The bonding region covers around 12 cm^2, which is 15% of the overall wafer. The lower die measures 12 × 12 mm^2 and includes bumps, redistribution layer (RDL), square interconnects and test pads. The upper die measures 10 × 10 mm^2 and has a dicing line, square interconnects, and bumps on its underside. A total of 2,700 bumps, each with a pitch of 20 μm and a size of 12 × 12 μm^2, are evenly placed around the periphery. A total of 27 test channels are created, using Kelvin and daisy-chain architectures. There are 120 square interconnects measuring 500 × 500-μm^2 that are embedded within the bumps in order to reduce the rate of failure as a result of wafer dicing. To rapidly transform Cu/Sn/Cu to a complete Cu/Cu_3Sn/Cu interface, two essential requirements must be met. The first criterion is that the ratio of hCu to hSn must be greater than or equal to 1.32, where hCu and hSn represent the combined thickness of Sn and Cu[19]. The second factor is the thinnest layer of Sn, which is mostly influenced by the pace at which the bonding process is heated. The temperature of the bonding was elevated from 27°C to 200°C in a span of 10 minutes in this study. The top wafer was constructed with Cu/Sn bumps that had a thickness of 3.0 μm for Sn and 5.0 μm for Cu. The lower wafer is equipped with copper bumps of uniform thickness. These bump structures facilitate the rapid removal of surplus tin by allowing it to react with the copper. The width of the Sn bump in the sandwich structure of Cu/Sn/Cu, which measured 12 micrometers, was made equal to the width of the Cu bump in order to simplify the manufacturing process.

4.3.4.2 Fabrication of Fine-Pitch Bump

The process for bumping the bottom and top wafers consists of a layer of SiO_2 with a thickness of 300 nanometers was produced as a passivation layer. The backside dicing line and RDL were created using wet-etching, sputtering, and patterning techniques, with aluminum layers that were 200 nm and 1 µm thick, respectively. The barrier layer was formed by sputtering a 50-nanometer thick layer of TiW, while the seed layer was formed by sputtering a 100-nanometer thick layer of Cu. Both layers were then patterned. First, copper bumps with a thickness ranging from 4.5 to 5.0 µm were electroplated. Then, tin bumps with a thickness ranging from 2.5 to 3.0 µm were electroplated, but only on the top wafer. Ultimately, the wet-etching procedure was utilized to eliminate the exterior barrier/seed layers. The spacing between bumps is 20 µm. The average thickness of the copper bump is 4.9 µm, while the tin bump has an average thickness of 2.9 µm. The thickness of the aluminum (Al) Redistribution Layer (RDL) is about 1.0 µm.

In order to determine the consistency of the bump, the standard deviation and mean bump height are employed. It happens when electroplating is going on. Bump uniformity is shown to be one of the key elements in making sure that pressure will be equally distributed during the bonding process. For each wafer, the bump height was measured at different locations (Table 4.3). The study revealed that standard variances of front and back wafers are equal to 0.19 µm and 0. That equated to homogeneous of 2.36% and 3.17%, respectively. The fact that the bonding process criterion was met was confirmed by these results.

4.3.4.3 Surface Pretreatment of the Bumps

Copper surfaces that have been electroplated don't fight oxidation well, and the thin copper oxide will make surface oxidation even faster. This phenomenon decreases the level of activity on the surface and hinders the diffusion of Cu-Sn, particularly in the case of SSD bonding. It was pretreated for 120 seconds with a plasma cleaning machine to get rid of the oxide layer and turn on the surface of the Cu bumps. The power used was 200 W, and the gas flow rate of Ar (5% H2) was set at 200 sccm. The XPS spectra demonstrates a significant decrease of 48.1% in the amount

TABLE 4.3
Bump Height[4]

Bump Location on Wafer	Top Wafer (µm) (Cu/Sn Bumps)	Bottom Wafer (µm) (Cu Bumps)
Middle	7.90	4.90
Up	8.30	5.20
Down	8.00	4.80
Left	8.20	5.10
Right	7.80	5.20
Average	8.04	5.04
Standard Variance	0.19	0.16

of oxygen on the surface, which decreased from 18.89% to 9.81%. Similarly, the angle of contact of deionized water decreased from 31° to as low as 4° (when completely extended), indicating that the surface of the bump was substantially active. Furthermore, the Sn surface was similarly affected by plasma pretreatment.

4.3.4.4 Bonding of Cu/Sn/Cu Using SSD

The SSD bonding process of Cu/Sn/Cu was carried out after pretreatment using a wafer bonder. The bonding was carried out at a pressure of 6.7 MPa, a temperature of 200°C, and a duration of 60 minutes. The process was conducted using a vacuum of 105 mbar. The bonded wafers were cut into individual dies. The cutting depth was controlled to remove the edges of the top die and disclose the test pads in the bottom die. There was no separation of the top die from the bottom die. The bonded dies were subjected to annealing at a temperature of 200°C for durations of 30 and 60 minutes in an environment of nitrogen gas in order to remove pure tin and the Cu_6Sn_5 phase.

4.3.5 Results and Discussion

The bonded dies' ability to join together was tested using five distinct tests, which encompassed electrical measurement, electromigration (EM) test, TC test, shear strength analysis, and interfacial analysis.

4.3.5.1 Interfacial Analysis

A scanning electron microscope (SEM) picture of the cross-section of a microstructure at the interface in its original bonded state depicts No overflow of Sn was seen, as occurred in Cu-Sn SLID bonding, due to the fact that Sn did not undergo melting at the time of bonding process at low temperatures. According to the findings, it was seen that Sn underwent a full transformation into intermetallic compounds (IMCs), specifically Cu_6Sn_5 and Cu_3Sn. The Cu_6Sn_5 phase, located in area 2, had a thickness of 0.85 μm and was positioned between the two Cu_3Sn phases (areas 1 and 3). The thicknesses of both Cu_3Sn phases were approximately 1.34 μm. The interface of the bonds were composed of five layers—namely, $Cu/Cu_3Sn/Cu_6Sn_5/Cu_3Sn/Cu$.

The SEM images which display the cross sections of microstructures annealing interface, respectively. 30 mins after annealing, the thickness of the Cu_6Sn_5 phase comes down to 0.49 μm, and the total thickness Cu_3Sn phase is 3.41 μm. Part of the Cu_6Sn_5 phase changed to Cu_3Sn. When annealing time is 60 minutes, the Cu_6Sn_5 phase was depleted, and the Cu_3Sn phase fused to form a 3-layer interface comprising $Cu/Cu_3Sn/Cu$. It contains a total of 3.66 μm Cu_3Sn phase thickness.

The pure Sn phase in this investigation was completely depleted throughout the bonding procedure. The Cu_6Sn_5 phase underwent consumption in order to provide a 3-layer interface that is more reliable during the successive process of annealing. The ultimate daisy-chain configuration is seen. Observation reveals the presence of Kirkendall voids between Cu and Cu_3Sn in both the annealed and as-bonded samples. This occurrence is because of the higher diffusion coefficient of Cu in contrast to Sn at 200°C. The existence of these voids has a notable impact on the dependability of Cu/Sn/Cu SSD bonding and necessitates investigations that are further required.

4.3.5.2 Analysis of Shear Strength

Die shear tests were conducted to assess the bonding strength using a bond tester model DAGE4000 made by Nordson DAGE.

The bonding strength of the as-bonded samples varied between 3.9 and 6.9 MPa, with an average of 5.0 MPa. The samples that underwent annealing at a temperature of 200°C for a duration of 60 minutes exhibited a maximum bonding strength of about 9.7 MPa and a minimum bonding strength of approximately 15.7 MPa. The mean value rose to 11.4 MPa. The fracture modes and surface composition were examined using energy-dispersive X-ray spectroscopy and SEM. Four models that are fractured were identified. Fracture in the as-bonded samples was limited to the Cu_6Sn_5 layer (Type I). In the case of annealed samples, the most common kind of fracture was observed in the Cu_3Sn layer (referred to as kind II), while additional fractures were found either along the Cu layer and TiW layer (referred to as Type III) or along the SiO_2 passivation layer and Al RDL (referred to as Type IV). The analysis revealed that either Cu_6Sn_5 or Cu_3Sn, both forms of IMC, would be the least strong layer in the Cu/Sn/Cu SSD bonding interface. The annealing process caused a significant transformation in the massive Cu_6Sn_5 grains, resulting in the formation of tiny Cu_3Sn grains with a columnar orientation. This transformation ultimately led to an increase in shear strength at the end.

4.3.5.3 Electrical Measurement

The Kelvin channel is composed of a single bonded bump and Al RDL. The daisy-chain channel consists of 100 interconnected bumps, an aluminum redistribution layer (RDL), and test pads. Theoretical resistance for materials annealed for 60 minutes is determined based on the appropriate resistivities, as presented in Table 4.4. The probing station was used to measure the electrical resistance (PE-4, EverBeing, Inc.). According to the data shown in Table 4.5, the resistance measured in Kelvin units, which varied between 35 and 40 m, was about double the expected theoretical value of 17.0 m. This discrepancy may be primarily attributable to the presence of flaws between the various layers. According to the data shown in Table 4.4, the channels were conductive and had a daisy-chain resistance ranging from 10 to 15, with an average of 12.2. This average value is comparable to the theoretical value of 9.6.

4.3.5.4 Thermal Cycling Test

The annealing samples underwent a 60-minute temperature cycling (TC) test ranging from 55°C to 125°C for a total of 1,000 cycles. The samples were exposed to these temperature extremes for a duration of 5 minutes each, as specified in the

TABLE 4.4
Theoretical Resistance[4]

Property	Bump(s)	A1 RDL	Test Pads	Total
Kelvin (mΩ)	2.8	14.2	—	17.0
Daisy-chain (Ω)	0.3	8.5	0.9	9.7

TABLE 4.5
Kelvin Resistance[4]

Samples	1	2	3	4
Kelvin Resistance in mΩ	36.6	39.0	37.4	33.3

TABLE 4.6
Resistance of the Daisy-CHAIN[4]

	TC of S1		TC of S2		TC of S3		TC of S4	
Property	Before	After	Before	After	Before	After	Before	After
Channel 1	11.6	12.9	13.5	13.8	11.6	11.9	10.3	12.4
Channel 2	12.9	13.4	13.1	13.7	12.9	13.6	10.9	12.5
Channel 3	13.3	13.7	13.1	12.8	12.9	13.6	11.4	13.4
Channel 4	14.3	13.4	12.5	11.4	11.7	14.1	10.5	12.6
Channel 5	14.9	13.8	12.2	11.5	11.6	14.5	10.3	12.2
Channel 6	15.0	13.7	11.3	10.5	12.1	11.7	9.6	11.2
Average	13.7	13.5	12.6	12.3	12.1	13.2	10.5	12.4
Total Average	12.2 (Before TC), 12.8 (After TC)							

JESD22-A104D standard, Condition B. The environmental conditions of the test chamber used for this purpose was the TSE-11 model manufactured by ESPEC, Inc. The reliability performance was assessed by identical experiments. The Table 4.6 shows that there were no noticeable defects, such as cracks or large voids in the bonded interface. The average bonding strength was approximately 13.3 MPa, little bit higher than the 11.4 MPa of the annealed samples. Additionally, the average daisy-chain resistance of 12.8 was similar to the 12.2 of the annealed samples. The annealed samples showed no noticeable change even after undergoing 1,000 cycles of TC testing.

4.3.5.5 Electromigration Test

Electromigration plays a crucial part in deciding the dependability of 3D interconnects, and its impact will intensify as the bump pitch decreases. In the conventional bonding process of copper- to tin-based solder, the failure of the electronic device was often due to either the presence of microcracks between Cu_6Sn_5 and Sn or the dissolution of the cathode Cu. These failures were primarily caused by the combined effects of current stress and chemical reactions. Through a crucial annealing procedure, the thin layer of Sn underwent a full transformation into Cu_3Sn. An enhanced structural stability of Cu_3Sn would significantly decrease the occurrence of chemical reactions during an electromagnetic (EM) test. The 60-minute annealing sample underwent an electron microscopy (EM) test at a temperature of 150°C using a current density of 5.0 × 10^4 A/cm^2 and 1.0 × 10^5 A/cm^2. The evaluation of the electromagnetic impact was conducted using daisy-chain resistance and interfacial microstructure. Following 140 hours of exposure to current stress of 5.0 × 104 A/cm^2, the resistance value, approximately ranging from 12 to 14, was almost identical to that of the annealed sample. During a subsequent period of 180 hours, a larger current stress of 1.0 × 10^5 A/cm^2 was applied, resulting in a noticeable rise in resistance. The ultimate resistance value reached a range of 80–100. The movement of few Cu atoms took place at the bonding contact between Cu and Cu_3Sn, resulting in the resistance rise. This Cu/Sn/Cu SSD bonding shows a superior capacity for anti-electromigration.

The bonding performance of Cu-Sn SSD bonding is mostly dependent on the diffusion between solid-state Sn and solid-state Cu, which is facilitated by the lower bonding temperature. Improved bonding quality might be attained by thoroughly cleaning and vigorously activating the bump surface, particularly the electroplated-Cu surface. The surface pretreatment significantly impacted the bonding of Cu-Sn SSD, and the time of bonding could be decreased by choosing an appropriate pretreatment technique and further refining the pretreatment parameters.

4.3.6 Conclusion

A wafer-level connection with a fine-pitch Cu/Sn/Cu structure was achieved utilizing SSD technology, with bonding occurring at a temperature of 200°C. The bonding performance was assessed using a variety of tests. The bonding contact did not experience any excess of Sn. The bonded interface was composed of layers of Cu, Cu_3Sn, Cu_6Sn_5, Cu_3Sn, and Cu. Subjecting the material to annealing at a temperature of 200°C for a time period of 60 minutes led to the formation of a three-layer boundary consisting of copper (Cu), copper-tin (Cu_3Sn), and copper (Cu). The results of the electrical resistance and die shear test calculation indicated that the Cu/Sn/Cu interface exhibited excellent conductivity and adequate bonding strength. There was no noticeable alteration in conductivity during the 1,000-cycle temperature cycling (TC) test, and it was also capable of withstanding a greater current load. Overall, Cu/Sn/Cu SSD bonding has the potential to be a technique employed in 3D interconnect for future 3D integration.[5]

4.4 LASER MICRO-WELDING

Laser micro-welding is a welding technique that forms tiny joins by harnessing the minimum input of heat of a laser, resulting in a tiny HAZ. The laser micro-welding joints, similar to conventional welding joints, consist of three components: weld metals (WM), fusion zone (FZ), and HAZ.[6] Micro-welding has extensive applications in the electronics, instrumentation, packaging, aerospace, and medical industries. Common uses include the connecting of thin sheet metal with a thickness of up to 0.5 mm, as well as wire or tubular components with a diameter of up to 1 mm. The Nd-yttrium aluminum garnet (YAG) laser has been increasingly utilized in electronic packaging to facilitate the connecting of smaller components. Lasers are ideal for micro-welding due to their ability to provide a precise quantity of energy to tiny components with a high level of accuracy.[7]

When it comes to combining different types of metals such as copper and stainless steel alloys, conventional procedures like crimping or press fitting have limitations in terms of preparation, repeatability, flexibility, and cycle time. Laser beam micro-welding has notable benefits, including a procedure that does not require physical contact and does not result in wear and tear. Additionally, it is a versatile method for connecting materials, as it allows for precise control over energy deposition, accommodates various component geometries, and works with different types of materials. Typically, steel and copper alloys are used for applications such as electrical connections and electromechanical components. The available space for the junction is typically limited, resulting in a corresponding limitation on the diameter of the beam. Only a pulsed laser source is appropriate because of its exceptional beam quality, which is twice as excellent as that of a continuous wave laser source at the same power level. Thus, the diameter of the beam 2w is halved at the same operating distance. Laser beam micro-welding has proven to be effective in joining several tiny components made of different metals in the watch industry.[7]

4.4.1 Challenges Involved in Micro-Welding

Material selection is a key design parameter in the design of micro-parts to avoid constraints on successful joining methods and subsequent assembly processes due to poor weldability of materials. Micro-parts have to perform a variety of functions, and for each function, there is a material that provides the best materials properties. The effect of the mechanical characteristics and its effects on weldability can be highly different due the geometries. However, micro-part design depends on more than just obtaining a weldable material. The effective design of micro-scale components hinges on the requirement for a given combination of materials and joint configuration. The weldability will depend on a variety of specific situations. For these cases, the effect of coupling energy and heat transfer on the scale could be very excessive as a result of bodily nature.

The absorptivity of a technical surface is a crucial issue, as it depends on several variables. The weld seam formation and power coupling are significantly affected by roughness, contaminations, and temperature. Based on the Fresnel equations, the absorbance of metals is also influenced by the incident angle and the polarization of the laser beam.

Stainless steel possesses favorable mechanical characteristics and is well suited for welding using laser beam. Compared to other metals, the absorptivity in the infrared (IR) to ultraviolet (UV) region is high. The austenitic stainless steel X5CrNi18 10 consists of 77% iron, 18% chromium, 5% nickel, and 0.05% carbon. K55 is frequently used to make connectors that are subjected to strong electrical or thermal stresses. The excellent electrical conductivity of copper is combined with the mechanical properties of an iron-based material in this copper alloy. Copper can only be welded with lasers at high intensities due to its high reflectivity. Once the copper reaches a liquid state, the pool of molten metal can become too hot, leading to the formation of welded splatters. The thermal conductivity facilitates rapid heat transmission inside the base of the material. This alloy of copper, known as $CuNi_3Si_1Mg$, consists of copper of about 96%, along with minor quantities of nickel (3%), magnesium (0.15%), and silicon (0.65%). Additionally, it has the intermetallic phase Ni_2Si. This alloy is specifically designed for applications such as laser welding. The weldability of brass in laser welding applications is reduced due to the fumigation of zinc and the subsequent generation of vapor. Zinc has a vaporization temperature of around 907°C that is less than the majority of other metals' melting point. When zinc is vaporized, it can lead to the expulsion of material, which in turn causes the formation of weld splatters, holes, and an uneven shape of the weld seam. CuZn37 consists of 37% zinc and 63% copper.

Many parameters are used to determine the quality of the weld: The primary objective is to create a secure joint when connecting two components. The tensile tests provide a reliable measure of the rupture strength, which is a valuable indicator for assessing the quality of the joint. Furthermore, it is imperative that the weld seam has a visually pleasing aspect. An essential factor for achieving a pleasing aesthetic is a uniformly smooth and level surface. The number of spatters and cracks must be minimal.[7]

4.4.2 Technique Used in Welding

The welding technique known as SHADOW utilizes a solitary Nd: YAG-pulse with a rapid speed of welding to produce a continuous seam of weld spanning many millimeters. Utilizing a high feed rate of 10 m/min and a single pulse of $\tau_H = 20$ ms, continuous weld seams measuring 3.3 mm were produced. Three distinct focal lengths were employed, yielding laser beam waist diameters of 100 µm, 200 µm, and 300 µm. A longer focal length lens offers the benefit of a greater working distance, which in turn helps to minimize contamination of the protective glass. The welding application required a power range of 250 W to 1400 W, which was enough. Typically, SHADOW leads to less energy input, resulting in less distortion and contamination of the component. The temperature of the molten pool is reduced, effecting a decrease in contamination of the workpiece by weld splatters or smoke. The energy is transferred by the process of conductive heat transfer. Steel materials are cooled because copper alloys have a greater thermal diffusivity. Due to the dissipation of heat, there is a reduced amount of energy available to sustain the molten steel pool. Materials shrinkage occurs when the metallic continuity is reached and the weld seam cools. Cold cracks will form if the residual stress exceeds the elastic limit. Typically, traditional approaches for predicting microstructure are not applicable in laser beam welding because of rapid rates of cooling and heating. There is not enough duration for any substantial dispersion to occur. Therefore, the development of intermetallic phases is also limited. The spatial arrangement of alloying components is closely linked to the movement of material inside the molten pool. Frequently, the melting zone has a turbulent structure, which arises from an inadequate mixing process at the time of laser pulse. The limited elements solubility leads to the creation of distinct, unmixed regions that are isolated by a thin layer of mixed zones. The dispersion of the various alloying components in laser beam welding is only partially regulated by diffusion. Dissimilar metal junctions exhibit compositional gradients and microstructural alterations, resulting in significant disparities in chemical and physical characteristics throughout the joint. The intricate arrangement of T-joints results in the distinction between t-laps and other parameters. Two distinct types of t-laps may be identified based on the direction of the laser beam: fillet weld and overlap weld.

The configuration of a T-joint can also vary. If the first component is made of steel, it is classified as NORMAL. If the copper alloy is component 1, then the arrangement is referred to as inverted. Consequently, there are eight distinct T-joints that may be identified (Table 4.7).

The angle at which a ray of light or other electromagnetic wave approaches a surface is referred to as the incident angle. Θ_i represents the angular separation between the rear piece and the laser beam. The feed angle Θ_s is determined by measuring the angle between the normal to the feed vector and the laser beam. An investigation has been conducted on welding methods that provide stinging sensations, as well as dragging operations. The distance between the joint interface and the front piece's optical axis of the laser beam is denoted by the laser beam alignment "d."

4.4.3 Results of the Experiments

4.4.3.1 Formation of Weld Seam

It has been possible to weld steel and brass, as well as steel and copper. The steel-copper combination is more weldable than the steel-brass combination in the majority of T-joint designs. Typically, the material copper exhibits a heat impacted zone in cross sections. The intermetallic phase Ni_2Si is absent in the material copper found in the melting zone, where it is often dissolved as $CuNi_3Si_1Mg$. Partial cold fractures have been discovered in the welding of standard copper TJFW. Typically, these cold fractures occur at the boundary between the molten puddle and the main body of steel. Then copper is used for cooling the steel material, causing a significant

TABLE 4.7
Configurations and Types of Weld Seams in T-Joints[7]

Notation	Component 1	Component 2	Weld Seam Type
NORMAL brass TJFW	S	B	FW
NORMAL copper TJFW	S	C	FW
INVERTED brass TJFW	B	S	FW
INVERTED copper TJFW	C	S	FW
NORMAL brass TJOW	S	B	OW
NORMAL copper TJOW	S	C	OW
INVERTED brass TJOW	B	S	OW
INVERTED copper TJOW	C	S	OW

S – Steel, B – Brass, C – Copper, FW – Fillet Weld, OW – Overlap Weld.

temperature differential that leads to residual stresses above the elastic limit of steel. The number of cold cracks mostly relied on the amount of power utilized.

When welding brass TJFW in an inverted position, there is a noticeable narrowing, but there is very little molten material on the backside. None of the heat impacted zone is seen in brass material. Nevertheless, the crystalline structure of brass has a very small grain size inside the region where melting occurs. Due to its low solubility in CuZn37, iron functions as a grain refiner and enhances the mechanical characteristics. The melting zone has a turbulent structure in all joint configurations. There are regions where very little blending has taken place. However, layers of overlapping areas also exist. A common structure is used in addition to mechanical mixing to create the metallic continuity.

A welding procedure that involves a stinging motion often yields somewhat superior outcomes compared to a dragging or perpendicular motion. Heating the components beforehand is one way to get this result.

In the case of vaporization, the majority of the energy of the laser beam is absorbed at the front end of the melt pool during the welding process, resulting in a stinging sensation. The melt pool's maximum temperature is also located in the anterior end of the laser beam. In the context of dragging processes, it is important to note that the anterior portion of the molten puddle is not melted directly by the laser beam but rather by the process of conductive heat. The drag process leads to a reduction in appearance, as well as a drop in weld penetration depth, resulting in a fall in rupture strength. The technical significance of this is currently being investigated.

The angle of incidence Θ_i in the fillet weld arrangement significantly affects the development of the weld seam. Altering the incident angle from $\Theta_i = 20°$ to $\Theta_i = 60°$ leads to a completely distinct seam of the weld.

In type I welding, the copper substance is completely melted over its whole thickness. The steel material has a melting zone of 150 µm. The solidification process does not exhibit any preferential orientation. In weld type II, the material steel is completely melted over its whole thickness. The steel substance undergoes cooling by contact with copper. Consequently, a greater amount of energy is transported to the side of copper, resulting in an asymmetric solidification zone. However, the shared length between the two components is significantly lower when employing an angle of incidence of $\Theta_i = 20°$. Currently, there are no suitable methods available to imitate the SHADOW welding of dissimilar metals. A much reduced or simplified perspective on absorption is a promising method to gain an understanding of this process. When the surface is in an optimal condition without any roughness or pollution, the laser beam shines on both components.

Both joint components absorb a portion of the laser beam's power based on their respective absorptivity. This process is referred to as primary absorption p. The power that is not absorbed by one joint partner is reflected by both plates. Therefore, a further absorption, known as secondary absorption, takes place. As a result of this dual absorption, the overall power coupling is enhanced.

The ratio of power P to twice the diameter of the beam 2w is a significant parameter for the shift from heat conduction welding to keyhole welding. Therefore, it should be possible to reduce the geometry to two dimensions. The use of the complete beam diameter as the reference value is not achievable due to the irradiation of both components by the beam. Four separate lines of relevance may be identified, assuming that a prospect reference value is a line on the material surface that is perpendicular to the feed vector: These lines are the major and secondary absorption lines of components 1 and 2, respectively. The beam diameter $2w$, angle of incidence Θ_i, and beam alignment d may all be used to calculate the length of these lines. The intensity of the laser beam decreases in proportion to the incident angle Θ_i since the surface is not perpendicular to the beam. Based on the geometric layout and Fresnel equations, we may define two ratios: the power absorption ratio $R_{P,\text{abs}}(1)$ and the consequent intensity ratio $R_{I,\text{res}}(2)$: The outcomes of the model using iron as component 1 and aluminum as component 2, which possesses similar optical and electrical characteristics like copper.

The Brewster effect is accountable for the occurrence of relative minimum and maximum values in the ratio of absorbed power, $R_{P,\text{abs}}$. Component 1 has up to eight times more power coupled than component 2. The resultant intensity ratio $(R_{I,\text{res}})$ is independent of the beam alignment (d) and increases as the angle of incidence (Θ_i) increases. Altering the incident angle from $\Theta_i = 20°$ to $\Theta_i = 60°$ leads to a reduction in the power absorption ratio from $R_{P,\text{abs}} = 7.6$ to $R_{P,\text{abs}} = 6.1$. Nevertheless, the ratio of the resultant intensity grows from $R_{I,\text{res}} = 2.9$ to $R_{I,\text{res}} = 10.7$. Furthermore, the heat conduction and power coupling exhibit distinct characteristics, leading to the process of vaporization and a specific type of welding known as keyhole welding. Based on experimental findings, the alignment of the laser beam does not affect this phenomenon, as long as its size is less than or equal to $d = 40\,\mu\text{m}$. This corresponds to the method described before. This phenomenon was also seen in inverted arrangements.

The morphology of the weld type I molten puddle closely resembles that of NORMAL copper TJFW. The presence of a robust amalgamation of steel and copper is readily apparent. A significant concentration of iron is dissolved in the copper substance. Moreover, copper has steel inclusions, but steel does not contain any copper inclusions. Type II weld seams have exceptional rupture strength. An inherent drawback of this welding technique is the complete penetration of the molten pool, leading to a significantly liquefied backside. The T-joints welded at an angle of incidence of $\Theta_i = 20°$ in the conventional fillet weld configuration exhibited greater rupture strength. The rupture strength can reach 310 N at standard copper TJFW and 270 N at standard brass TJFW.

4.4.3.2 Copper T-Joints

Conventional TJFW results in more robust and aesthetically pleasing weld seams compared to INVERTED TJFW. Standard copper TJFW usually has a smooth surface, which results in a very pleasing weld seam appearance. The maximum strength of rupture is 320 N. In the INVERTED copper TJFW, there is a higher concentration of copper in the joint region. Increased laser power is required to achieve the molten puddle due to copper's poorer reflectivity and increased heat conductivity. The weldability is worse compared to the NORMAL setup. The maximum strength of rupture is limited to 180 N. The visual aesthetics are compromised in the NORMAL configuration due to the presence of a melted backside of the copper substance. There is a minimal presence of weld splatters seen for standard copper TJOW. The steel back piece consistently produces smooth and regular weld seams of excellent aesthetic quality, except when subjected to high laser powers. In addition, the level of smoke is often lower compared to TJFW. The strength of rupture of a TJOW is somewhat lower, measuring at 270 N, when compared to the TJFW. Similar to other types of welded joints, there is a depression present at the end of the weld seam in TJOW. The depression mostly results from shear stress induced by temperature variations on the top side of the molten puddle. Within the molten puddle, there is a decrease in pressure near the center, while a raise in pressure is observed along the border. The rise in pressure occurs due to the deceleration of the molten material flow at the boundary with the material base. This may be prevented by using the correct pulse shaping technique.

The laser power constraints of INVERTED copper TJOW are significantly more limited compared to NORMAL copper TJOW. The discrepancy between the bottom and higher limits of laser power is approximately 15% for a focal length of $f = 90$ mm. The disparity for standard copper TJOW is around 50%. Upon reaching a molten state, the copper exhibits an enhanced absorptivity while experiencing a reduction in conductivity of heat. The molten puddle is at risk of overheating. The quantity of smoke is similar to that of a typical copper TJOW. The maximum strength of rupture is 260 N.

4.4.3.3 Brass T-Joints

Typically, the visual quality of NORMAL brass TJFW is inferior to that of NORMAL copper TJFW. The primary cause is the fumigation of zinc, which leads to an unstable laser welding process. The tensile strength is diminished and the presence of voids and emission of smoke is substantially elevated. When welding standard brass TJFW, the maximum rupture strength is 260 N. The weld seam exhibits a satisfactory look. When working with inverted brass welding, it is common to encounter the production of greater quantity of smoke and holes. Additionally, the results from the experiments exhibit a significant amount of variation. The rupture strength is relatively low, measuring around 150 N. It is highly advised to avoid this joint configuration due to its poor weldability.

The visual quality of a typical brass TJOW is almost on par with that of welding copper TJOW. In both situations, steel serves as the base component, and the energy transfer takes place within the same material. In the material brass, there are multiple incorporations of steel grains present in the melt pool. Metallic fingers enclose the brass plate, creating a toothed connection between the different metals. In reference to the additional steel-brass weld connections, there is a noticeable narrowing in the brass material close to the interface of the joint. The strength during rupture measures approximately 200 N.

The laser power restrictions of INVERTED brass TJOW are lower than those of INVERTED copper TJOW because of the stronger fumigation of zinc and its increased absorption capacity. In general, the rupture strength values tend to be lower compared to copper TJOW. The incomplete weld seam length is highest in the inverted brass TJOW compared to the other three types. There appears to be a reduction of the material on the upper side of the brass. Smoke is generated in significant amounts for almost every combination of parameters. The values for the appearance are quite low. Of all the T-joint designs that were looked at, the inverted brass TJOW has the most unsightly weld seams. Nevertheless, the strength of the rupture with 220 N is remarkably excessive. To improve the welding procedure of this joint arrangement, more research is required.

4.5 MICRO-FRICTION STIR WELDING FOR MULTI-HYBRID STRUCTURES

Micro-friction stir welding is employed to join thin electrical components and micro-mechanical components, commonly found in industrial applications. The speed of rotation of the tool, together with the welding speed and angle of tool tilting, play a crucial role in the μFSW process and have a significant impact on the properties of the joint. It provides superior joint properties compared to the properties of the base material.[8]

4.5.1 Introduction

The micro-joining technology presents a fascinating and complex challenge in today's modern world. There has been a growing need for micro-joining technologies to make modern technological devices lighter and smaller. Welding soft and light materials offers a significant challenge to contemporary welding technology. When it comes to manufacturing small parts, friction stir welding (FSW) has proven to be a highly effective process for creating cost-effective parts and achieving strong joints in assemblies. In addition, FSW is recognized as an environmentally friendly welding technology due to its low energy consumption and absence of gas emissions during the welding process. Since consumable resources like filler materials, shielding gases, fluxes, and electrodes are not required, it is also regarded as an economical and environmentally beneficial welding technique. A survey conducted in 2002 by the American Welding Society (AWS) revealed a significant expenditure on arc welding equipment in the United States. Due to its increasing popularity, FSW has quickly become the preferred welding process, replacing conventional methods. With the telecommunication industry expanding, automotive sensors on the rise, and advancements in the medical field, there is a growing demand for joining smaller conductive materials. With the ongoing trend of miniaturization, there is an increasing need for innovative micro-welding techniques. Traditional methods like soldering and brazing are becoming less effective due to issues such as high joint resistance and the need for reliable and long-lasting joints.

μFSW is capable of joining thin materials, such as micro-mechanical components, micro electrical components and electronic components. Over 27 years, significant achievements have been made in the area of FSW. Numerous journals have reported on the FSW joining process of dissimilar and similar materials and related developments.

However, when it comes to micro-FSW (μFSW), there have been limited publications exploring various aspects of this topic. Thanks to the continuous progress of technology, μFSW is now utilized for joining various innovative materials. This section provides a concise overview of the μFSW process, covering various aspects and various materials. This assessment is accompanied by a thorough assessment of the key challenges faced in the μFSW process, including microstructural behavior and mechanical characteristics. Lastly, conclusions are made with an eye toward upcoming problems and the future course of study.

4.5.2 An Overview of μFSW

In several sectors, joining techniques that are solid state in nature are becoming increasingly used for combining softer materials that are difficult to fuse using traditional welding methods. FSW was created as a method of joining materials without melting them, and it was devised and perfected at The Welding Institute (TWI) in the United Kingdom in 1991. While it was actually developed and created specifically for Al-alloys, it is now utilized for connecting a wide range of metal matrixes of soft and hard metals and composites. Over 27 years, this invention has achieved a significant victory in combining high-strength aluminum alloys, magnesium alloys, copper alloys, titanium alloys, and metal matrices used in marine, aerospace, defense, shipbuilding, and automobile industries.

In 2004, Nishihara et al. successfully used the FSW method to join plates made of magnesium alloy AZ31 that are 1000 μm or thinner. The authors dubbed the altered methodology of FSW as μFSW technology at that time. Micro-friction stir welding is based on the same idea as friction stir welding. Some researchers utilized μFSW to connect wooden pieces and demonstrated that the strength of the weld was reduced as a result of the restricted depth of the weld line. The optimization of weld settings is deemed crucial. Some utilized μFSW to connect different alloys of aluminum—namely, AA6082-T6 and AA 2024-T3 sheets with a thickness of 0.8 mm. The author said that weld joints exhibited outstanding mechanical characteristics, and the tensile failure, disregarding the existence of flaws, was attributed to the variable thickness of the shoulder plunge depth occurring at the nugget zone (NZ). In 2012, Sattari and his team employed a butt welding joint to join 0.8 mm thick AA5083 alloy using μFSW. They successfully produced defect-free samples by conducting the welding process at temperatures varying from 430°C to 510°C. Microscopic flaws are produced in the NZ as a result of the high welding speed. The study conducted showed that the use of μFSW to join thin aluminum plates resulted in joints with high flexural strength, excellent ductility, and favorable microstructure. The joints were also subjected to testing, which revealed that they could withstand bending without breaking up to 180°. Micro-friction stir welding is employed to connect thin-walled constructions, electronics, and electrical components; micro-mechanical assemblies are often found in many industries. In 2010, Wang et al. successfully joined very slim sheets of different metals using μFSW, which is regarded as a successful approach for combining micro gadgets, the medical industry, and micro-electronic components. An investigation was conducted on the failure of 300 μm ultra-thin aluminum 1,050 sheets. The failure originated in the stir zone at the tip of the notch and spread down the NZ, ultimately leading to the final failure. The formation of cracks occurs in the HAZ and the thermomechanically affected zone (TMAZ), resulting in the fracture of the welded specimen. The application of μFSW has been the subject of extensive investigation to connect various materials, including extremely thin and ultra-thin sheets of copper alloys, titanium alloys, aluminum alloys, other dissimilar alloys, and wood.

However, the majority of the study has focused on the application of μFSW on aluminum sheets. In their study, some researchers effectively conducted welding utilizing 1 mm thick plates of copper DHP. The purpose of their experiment was to demonstrate the significance of process factors in the joining of alloys. In 2015, Papaefthymiou and his team achieved effective bonding of ZnTiCu sheets with a thickness of 700 μm using a μFSW technique. The resulting joint exhibited excellent integrity and improved formability when carrying out the welding at a traverse speed of 318 mm/min and a rotational speed of 1000 rpm. In their study, in the year 2017, Huang et al. demonstrated the successful joining of a 0.5 mm thick aluminum sheet—namely, Al-6061—utilizing a μFSW technique. However, a plunge depth of 0.05 mm results in an empty groove, making it easier for larger flashes and welding errors to occur. The use of μFSW presents a wide range of opportunities and difficulties in achieving strong joints when combining thin sheets for various applications. S Shankar et al. examined the combined properties of a friction stir welded joint between aluminum and copper.

4.5.3 Process Involved in Micro-Friction Stir Welding

The μFSW revolutionized the process of connecting metal alloys, materials with low melting points, and softer metals. The technique involves employing a rotating μFSW tool that is nonconsumable, consisting of a profiled pin onto a shoulder. The sort of welding connection and the characteristics of the base materials to be welded determine the pin profile geometry and μFSW tool material to use. Multiple researchers have recorded the different FSW tool designs, materials, and tool pin profiles and strengths. The μFSW process is regulated using two methods: displacement control and pressure control.

The procedure consists of two sides: the "advancing side" and the "retreating side." The μFSW procedure is broken into four steps. The four phases involved in the μFSW process are described as follows:

During the initial phase of μFSW, a cylindrical tool with a specifically modeled profiled pin is introduced into the beginning of the joint line between the plates. The plates are securely fastened with a rear support bar positioned under the weld line. The tool is made to rotate at a consistent velocity and descends with a vertically applied force until the tool's shoulder makes contact with the surface of the work item. This process is referred to as the "plunging" process. The process is achieved by the combined effects of frictional heat production and changes in the material's condition resulting from the rotation of a tool, resulting in mechanical deformation. Friction taking place among the surfaces of work and revolving tool components develops heat in this process. The shoulder of the μFSW tool experiences friction with the surfaces of the workpiece. The friction between the surface of the plate and the tool causes an increase in temperature. This elevated temperature causes the metal beneath the tool to enter a plasticized condition, allowing it to undergo mechanical deformation. By inserting the pin into the substance, friction occurs, resulting in the generation of heat. This heat causes the material to become soft and fluid, allowing it to be stirred.

Following the successful completion of the "plunging" operation, the "dwelling" operation begins. During this operation, the material is made to flow from the main edge to the trailing edge of the spinning tool. The downward push exerted by the shoulder helps to contain the evacuation of softer material. Currently, a semiliquid substance moves along the revolving tool and undergoes significant amounts of plastic deformation. During the third stage of the μFSW process, a tool that is rotating moves steadily through the weld line at a consistent speed. The mixing together of fluidlike materials takes place as a result of the continuous axial force applied by the tool. Material mixing occurs by either bonding or atomic diffusion, which is determined by the temperature and pressure during the tool passage. During the last stage, the process of retracting or cooling is performed. Once the final destination is reached, the tool is withdrawn from the workpiece, resulting in the creation of an exit hole. Various techniques are employed to remove the exit hole that is created at the conclusion of the μFSW process in the intended weld by filling it. Occasionally, the length of a weld exceeds the intended length.

4.5.4 μFSW Process Parameters

Force, torque, tool traverse speed, tool inclination angle, depth of shoulder plunge, and tool geometry are the important parameters. Downforce in μFSW is a significant force needed to achieve proper contact between the surfaces of the workpiece and the tool shoulder, with the probe penetrating correctly. Friction downforce is developed by applying pressure between the tool shoulder and the surfaces of the workpiece, resulting in increased load.

The application of torque to the tool generates friction at the shoulder of the tool. The force of rotation generates interior friction between the weld material and the tool probe within the area where the welded materials are plastic. It also makes it easier for the weld material to flow through the tool pin profile. The primary purposes of tool pin shape and tool tilt angle are to manage material flow and mitigate the generation of frictional heat across successive generations. Tool geometry is designed to maintain the materials in a plastic condition beneath the shoulder of the tool. The input power from the μFSW machine, which refers to the energy per unit length at traverse velocity, is a crucial element in the μFSW welding process.

4.5.4.1 Selection of Tool in μFSW

A nonconsumable tool that rotates and is utilized in the μFSW process. The object consists of two primary components: the tool shoulder and the tool pin. The key variables that affect the results of the welding process are the diameter of the tool shoulder, the diameter of the pin, the length of the pin, and the pin profile. The pin and tool shoulder are utilized to generate heat for plasticizing the workpiece at the joining zone, regulate the flow of material, and contain molten metal beneath the shoulder. Typically, the shoulder of the tool produces 80% of the overall heat, while the tool pin creates the remaining 20%. The length and size of the pin and shoulder are crucial to this element. During the insertion of the tool (tool plunge), the pin is placed into the space that is below the tool shoulder, causing displacement of the material.

In their study, during the year 2013, Galvao et al. examined the properties of the weld by considering three different geometries of the shoulder: scrolled-shaped, conical-shaped and flat-shaped. They find that the flat-shaped shoulder is not suitable for creating a joint, whereas the scrolled and conical shoulders are more suitable for achieving a finer grain and higher strength structure. These desirable properties can be achieved by using a minimum rotational speed.

The length of the tool pin is slightly smaller than the thickness of the workpiece, while the tool diameter is somewhat greater than the thickness of the workpiece. The tool pin has threads that are utilized to exert a strong force on materials, driving them into the chamber of the tool shoulder from the lower end of the pin. This helps to force the materials that are already there into the pin's flow. The μFSW technique utilizes a range of tool pin profiles, including hexagonal, cylindrical, square, triangular, tapered pins with cylindrical form, and three flat tapered pins.

In their study, during the year 2018, Hussain et al. employed five different shapes of tool pin profiles: tapered cylindrical, cylindrical, threaded cylindrical, triangular, square, and hexagonal. They found that all five tools were suitable for welding aluminum alloys. Among these, the tapered cylindrical profile tool yielded the tensile strength with the highest value in the weld joint. On the other hand, the triangular tool resulted in a lower tensile strength value for the weld joint. The weld connection exhibits its highest impact strength when a tapered cylindrical tool is used. The use of a conical taper profile tool at 1400 rpm can result in weld joints that are free from flaws and have outstanding mechanical and microstructural qualities. To achieve optimal welding properties, it is recommended to maintain the tool's inclined angle, also known as the tilt angle, between 2° and 4° relative to the normal of the workpiece.

4.5.4.2 Microstructure of μFSW Joint

The horizontal cross-section of μFSW reveals four distinct zones: the NZ, the TMAZ, the HAZ, and the base metal (BM). The grain boundaries, grain size and metallurgical characteristics of these four different zones differ from those of the BM. The NZ is created at the interface between the base of the metal and the tool pin. The stirring action at NZ is achieved by adding heat and high-strain energy to the shoulder and tool pin. The size of the grain at the NZ is smaller than the size of the grain at the BM due to dynamic recrystallization. Smaller sizes of grains enhance the strength and microhardness of the stir zone in the weld.

The area of the TMAZ experiences heat and undergoes minimal distortion due to insufficient plastic strain. In the welding zone, the BM grains are consumed in an upward direction relative to the material flow. In μFSW, only heat is delivered to the HAZ, and the effects are negligible. In the BM, the microstructural and mechanical changes are minimal and can be considered insignificant in μFSW. In the year 2015, Papaefthymiou et al. found that the joining area included onion ring patterns, but no microstructural alterations were seen in the HAZ region. The use of threaded pins resulted in the formation of an onion ring pattern. In the year 2017, Huang et al. said that the width of the NZ, TMAZ, and HAZ increases as the velocity of rotation increases. They further clarify that a fine grain structure may be observed in the NZ, which enhances the strength and hardness of the weld.

4.5.4.3 The μFSW Joint's Mechanical Properties

The μFSW effectively resolves many complications and imperfections such as porosity, blowholes, and additional problems relating to the solidification of molten metal and the dissolution of gases during welding. Frictional and mechanical work create heat, which is then transformed into heat energy, causing a rise in the temperature of the workpiece. Increasing the temperature results in a decrease in the strength of the work item. Due to the elevated temperature during the welding process, a new grain is generated in the welded zone as it surpasses the recrystallization temperature. The increase in temperature and compressive loading results in the formation of smaller and uniform grains, hence enhancing the weld's strength. In the year 2012, Sattari et al. said that a decrease in the rate of speed leads to a rise in tiny flaws. They found that for Al-5083 with a thickness of 0.8 mm, flawless welding occurs between temperatures of 430°C and 530°C. In the year 2017, Huang et al. demonstrated that when attaching Al-6061 with a thickness of 0.5 mm using a pin with three flats, the resulting tensile strength is 90% of the BM. Enhancing the rotational speed enhances the surface characteristics of the weld and removes the presence of the kissing bond fault.

4.5.5 Multilayer Aluminum Alloy Sheets

Using the μFSW method, annealed AA1100-O sheets with a thickness of 0.254 mm were joined into very thin sheets. Customized spinning tools, built from high-speed steel rods, were employed for this purpose. The microstructure of the sheets in their original state, having a grain size of around 25.5 μm on average. During the cold rolling process to achieve the final gauge, the sheet surface experienced higher deformation relative to the central region of the as-received material. As a result, the grain development during the subsequent annealing process was accelerated. Consequently, a notable disparity in grain size was detected, with the grains adjacent to the sheet surface being bigger than those in the central region.

The Charly 4U, a four-layer design of the multilayer joints was welded using a computer numerically controlled milling machine. The configuration is, where there is a top-to-bottom sequence of layer placements, with the top layer being identified as the first layer and the bottom layer as the fourth layer. The welding direction was kept parallel to the direction of sheet rolling throughout the μFSW process. The study focused on investigating four essential welding factors, each with three levels, in order to examine the process window. Utilizing a created L9 orthogonal array with nine runs, the Taguchi optimization technique was employed to reduce the quantity of experimental specimens. In order to address the limitation of soft alloys with low surface friction coefficient, large speed of rotation (e.g., 10,000 rpm) were employed to increase generation of heat and improve weldability. The impacts of tool shape were evaluated using three tools without threads with distinct geometries of the pin: straight cylinder pin, tapered cylinder pin and stepped pin. The three pins had a uniform height of 0.8 mm. Additionally, a curved chamfer with a radius of 0.35 mm was incorporated into the design of the shoulder. This chamfer was intended to avoid the accumulation of material chips during the μFSW process.

An optical microscope was used to visually inspect all the multilayer joints and evaluate the surface characteristics. Electron backscatter diffraction (EBSD) was employed to analyze the microstructure of defect-free multilayer joints that were achieved by following the anticipated ideal welding state. In order to get the EBSD samples ready, a precise diamond saw operated at low speed was used to remove small portions from the joints in a transverse manner. Subsequently, these sections underwent a mechanical polishing procedure. A Vickers hardness indenter was used to assess the microhardness of the defect-free joints with several layers. The indenter was loaded to 100 grams, and the dwell time was set to 15 seconds. To ascertain the hardness distribution, a microhardness measurement was carried out throughout the whole weld zone and in the middle of each layer. For the weld zone and the remaining area, the measurement intervals were chosen at 0.5 mm and 1 mm, respectively. Three duplicate measurements of each of the four layers' microhardness values were made.

In order to assess the mechanical characteristics of the joints and enhance the welding process, the highest amount of tensile force that the complete four-layer joint design can withstand was calculated. Under projected optimal welding circumstances, changes in bonding strength in the defect-free joints of the multilayer structure were examined using quasi-static tensile testing on each layer and peel tests on both the retreating and advancing sides. In the layer-wise tensile testing, just the layer being tested was secured at both extremities while some layers bent at a 90° angle. In testing of peeling, just the two layers next to each other were secured in order to measure the highest amount of force required to separate them and assess the strength of the bond. As a result, three different peel configurations were obtained for the joints consisting of four layers (namely P12, P23, and P34, where Pxy depicts the arrangement used for the peel test between the x- and y-th layers). The samples utilized were rectangular in shape and had dimensions of 60 mm × 6 mm × 1.016 mm. With a low-speed precision diamond saw, they were cut in the joint's transverse direction. The tensile and peel tests were carried out in compliance with ISO 10447 (2015) using an Instron testing apparatus at room temperature and a constant crosshead speed of 2 mm/min. For every welding situation or peel configuration, there were three iterations.

4.5.6 Results and Discussions

The section presents the establishment of the procedure window to achieve connections free from defects with many layers using μFSW. The study takes into account four key welding factors, which include the tool geometry and three welding parameters. The joints without any defects are next assessed regarding their bonding strength, microhardness, and microstructure using the anticipated optimal conditions for welding.

4.5.6.1 Process Window Creation

The multilayer joints' surface morphology in accordance with the L9 orthogonal array. There are four different surface morphologies that have been seen. These are surface flash (observed in trials #1, #2, #3, and #7), irregularly shaped semicircular hooks (observed in trials #3 and #6), surface-open tunnel (observed in trials #5 and #9), and smooth surface (observed in trials #4 and #8).

Insufficient heat created throughout the process may not be enough to make the material next to the spinning tool soft. However, the friction exerted between the workpiece and the tool shoulder might still create significant shear stress, leading to the formation of flash on the surface. The notable flash noticed in the initial three trials, conducted at an identical rotation rate of 10,000 rpm, suggests that there is an inadequate production of heat at this particular rate of rotation. Extruded flash may also form with increased rotation rate if a very large tool plunge depth is used. For example, during trial #7, the tool rotates at a rate of 30,000 rpm, creating enough heat to fully mix the top layer due to the smooth contact surface. However, the excessive plunge depth of 0.9 mm results in the expulsion of an extreme quantity of flash.

When the material agitated is not sufficiently softened by the friction heat, it can lead to the formation of an irregular welding path with discontinuous semicircular hooks. This happens when the advancing and retreating sides experience nonuniform shear stress distribution and material hardening. An example of this can be seen in trial #6. Trial #5 and trial #9 exhibit notable surface-open tunnels, while having the identical tool geometry of a straight cylinder pin. This suggests that the amount of substance being agitated is inadequate to cover the whole welding route with this specific tool geometry. Runs #4 and #8 have clean joint surfaces, indicating that the top layer has been blended well during welding under both of these settings.

By visually examining the surface characteristics of the joints, it may be inferred that there is a limited range of conditions that must be met in order to achieve joints without any defects. A high rotational velocity, such as 20,000 revolutions per minute (rpm), is essential for producing adequate frictional heat. The quality of the joint is highly dependent on the depth at which the tool is plunged. This depth is determined by finding the right balance between the quantity of material that has to be agitated and the occurrence of flash. The quality of the joint is also significantly influenced by the spinning tool's shape and velocity. It is important to generate enough heat to avoid the formation of incomplete semicircular tunnel and hooks faults.

In order to enhance the joining process and assess the impact of four specific welding factors, the mechanical characteristics of the joints are analyzed under various welding situations. In the Taguchi optimization approach, signal-to-noise (SN) ratios derived from experimental data are frequently utilized to evaluate joint quality. As defined by Roy in the year 2001, the signal represents the intended real values, while the noise represents the undesirable variables in the measured values. Using the larger-the-better (LB) criterion, the joint's tensile load is maximized. This criterion is based on the SN ratio, which may be expressed as

$$S/N = -10\log\left(\frac{1}{n}\sum_{i=1}^{n}\frac{1}{y_i^2}\right). \tag{4.1}$$

The number of duplicates for each welding situation is indicated by the variable n, with a value of 3. The variable y_i represents the highest tensile force for the i-th specimen.

TABLE 4.8
The L9 Orthogonal Array's Maximum Tensile Loads[9]

	Maximum Tensile Load (N)				
Run#	1st Specimen	2nd Specimen	3rd Specimen	Average	S/N Ratio with LB Criterion
1.	117.6	120.3	125.5	121.1 ± 3.3	41.7
2.	129.9	120.7	105.3	118.6 ± 10.2	41.4
3.	187.4	189.7	188.8	188.6 ± 0.9	45.5
4.	310.8	305.5	290.6	302.3 ± 8.5	49.6
5.	178.5	184.8	199.6	187.6 ± 8.8	45.4
6.	151.1	171.3	176.0	166.2 ± 10.8	44.4
7.	149.2	122.0	135.8	135.7 ± 11.1	42.6
8.	281.8	283.2	292.9	286.0 ± 4.9	49.1
9.	190.9	189.6	189.7	190.1 ± 0.6	45.6

Table 4.8 presents the highest amounts of tensile stresses for the whole multilayer joints and the SN ratios, as determined using the L9 orthogonal array. Research has shown that joints without exterior defects have the highest load-bearing capacity due to the establishment of robust bonding. As an example, the highest amount of force that can be applied without breaking in trial #4 is 302.3 N. Additional faulty joints demonstrate reduced load-bearing capacities.

Considering the LB criteria, a greater SN ratio indicates that a factor has a more substantial impact on raising the mean response, which in this case is the greatest average tensile load. The average maximum tensile load shows a considerable initial improvement as the rotation rate increases, followed by a reduction. The ejection of the softened stirred material at greater rotation speeds is the cause of this drop. The slight fluctuations in the SN ratios, with a delta values of 1.0 for plunge depth and 0.7 for welding speed, suggest that these factors have a very little impact on joint quality compared to the tool geometry and rotation rate.

The joints welded using stepped pins have a higher load-carrying capability than joints welded with other kinds of tools. As per the modeling analysis, the production of heat in μFSW may be divided into three components: The following can be used to depict the heat produced by the pin tip surface (Q_3), pin side surface (Q_2), and shoulder (Q_1):

$$Q_1 = \frac{2}{3}\pi\tau\omega\left(R_{\text{shoulder}}^3 - R_{\text{pin}}^3\right)(1+\tan\alpha) \tag{4.2}$$

$$Q_2 = 2\pi\tau\omega R_{\text{pin}}^2 H_{\text{pin}} \quad \text{for cylindrical surface} \tag{4.3}$$

$$Q_2 = \frac{2}{3}\pi\tau\omega\left(R_{\text{top}}^2 + R_{\text{bottom}}^2 + R_{\text{top}}R_{\text{bottom}}\right)H_{\text{pin}} \quad \text{for tapered surface} \tag{4.4}$$

$$Q_3 = \frac{2}{3}\pi\tau\omega R_{\text{pin}}^3 \tag{4.5}$$

The shear stress at contact, denoted as τ, is assumed to be consistent. ω represents the rate of rotation, while R_{pin} and R_{shoulder} refer to the radii of the tool pin and shoulder, respectively. α represents the angle of tilt, and H_{pin} denotes the height of the pin. The angle of tilt α in this investigation is zero. The friction coefficient and contact pressures both affect the contact shear stress, and accurately determining these factors throughout the stirring process can be challenging. In order to streamline the impact of tool shape on heat generation, it is assumed that the uniform contact shear stress remains constant. The heat production and the specific areas where heat is generated for each of the three rotating tools. It is important to observe that an extra heat source (Q_4) is indicated for the stepped pin. The heat produced by the friction between the workpiece and the pin step is matched by this extra heat source. The heat values produced by the three devices in their respective scenarios are presented in Table 4.8. These values were computed using equations (4.2)–(4.5). The analysis reveals that the stepped pin generates the highest amount of heat $(1.672\pi\tau\omega)$, while the tapered cylinder pin generates the lowest amount $(1.560\pi\tau\omega)$. These findings align with the SN ratios. It is assumed that generating a greater quantity of heat can enhance the properties of the joint. It is noted that the primary method of heat creation in the μFSW process is through friction at the surface between the workpiece and the shoulder. The friction caused by the tip of the pin, on the other hand, generates a little amount of heat. Regarding the stepped pin, the interaction between the step of the pin and the workpiece results in the production of extra heat. This heat is more substantial than the heat created by the friction between the tip of the pin and is advantageous for creating a strong bond in the lower layer.

Based on the data provided in the previous research works, the most effective welding condition for achieving the highest load-carrying capability can be determined. For this, a stepped pin must be used, along with a rotation rate of 20,000 rpm, a welding speed of 3 mm/s, and a plunge depth of 0.85 mm. By examining the fluctuations in the SN ratios presented in Table 4.5, under the tested conditions, we can ascertain how these welding parameters affect the maximum tensile load. The ranking of their contribution is as follows: The sequence of factors in the welding process is as follows: tool geometry, rotation rate, plunge depth, and welding speed.

4.5.6.2 Characterization

The microstructure's cross-section of a standard defect-free junction with many layers, achieved by implementing the anticipated ideal welding parameters. All four layers are flawlessly joined, as shown by the cross-sectional microstructure. Furthermore, the geometry of the weld zone is inverted, tapering down from the top layer to the bottom layer in terms of the length of bonding between neighboring layers. As an illustration, L12 is 1.52 mm, L23 is 1.25 mm, and L34 is 0.81 mm in length. The joint's overall thickness, measured along the weld center, is 1.01 mm, which is almost the same as the combined thickness of the four stacked layers of base material, which is 1.016 mm. However, there is a significant initial layer thinned out, particularly on the withdrawing side. The base material's thickness drops from 0.254 mm to 0.21 mm, resulting in a 17.3% thinning.

The microstructures on the retreating and advancing sides: The second layer's enlarged microstructures on both sides—designated as "2" and "3" In order to examine the variation in microstructure, the microstructure of the dynamically recrystallized zone—also referred to as the stirred zone (SZ) and designated as "1" The study reveals a remarkable refinement of grains in the SZ, with the typical grain size decreasing from 25.5 μm (the size of grains in the base material) to 2.01 μm. On the retreating side, there is a distinct transition zone called the TMAZ between the base material and the SZ. This zone is distinguished by a severely distorted structure that lacks recrystallization. The TMAZ width diminishes from the uppermost layer to the lowest layer. In comparison to the retreating side, the advancing side displays a greater shear stress gradient. As a consequence, a distinct boundary between the SZ and the base material is formed. This is because solid Al on the advancing side transforms into softened Al. Furthermore, there is a lack of noticeable expansion in the grain size in the TMAZ region, specifically it is difficult to identify the HAZ that experiences a thermal cycle without experiencing any plastic deformation.

With the base material's hardness provided as a reference, the microhardness distributes in the four layers. The origin point is the weld center, and the retreating and advancing sides are represented by the left and right sides of the coordinate system, respectively. There is no substantial decrease in hardness (i.e., in comparison to the basic material, the microhardness value is not less). This suggests that the dimensions of the HAZ are too minute to be detected using microhardness testing, which aligns with the findings from the microstructure analysis. Due to the ample amount of heating due to friction and extensive plastic deformation that occurs during the μFSW process, there is a noticeable increase in material hardness in the weld zone. The first layer has the biggest hardened area, as can be seen by comparing the microhardness distributions in the other layers. This is due to the fact that it is directly in contact with the surface of the shoulder and experiences significant friction. Furthermore, the weld center's microhardness value for each of the four layers is around 33 H V. In addition, the breadth of the toughened area on the side that is moving away (~1.5 mm) is determined to be greater than that on the side that is moving forward (~1.0 mm).

The average maximum tensile load for the joints collectively in the tensile test is 366.1 N, which is over four times the load-bearing capacity of the base material. The fracture occurs in the base material instead of the weld zone, demonstrating that the base material's tensile strength is less than the cohesive strength of the entire joint. Layer-wise tensile testing revealed that each layer experienced breakage originating from the base material. According to these occurrences, each layer's bonding strength is probably stronger than the underlying material's tensile strength. Consequently, it is challenging to distinguish between the variations in these layers' bonding strength and the tensile test outcomes.

In order to examine the fluctuation in strength of the bonds inside a defect-free joint consisting of many layers, tests for peeling are conducted between the neighboring layers. The highest possible peel load is utilized as a measure to assess the bonding strength. The advancing and retreating sides have different material flows, which might lead to the formation of different bonding strengths. Peel tests are thus carried out on both surfaces. To assess the disparity from the weld zone, a review is conducted on the base material's maximum tensile stress and fracture morphology. The tensile load-displacement curves of the corresponding fracture surface and the base material are shown in (b) and (a), respectively. While there may be differences in the displacements leading to fracture, the maximum tensile loads are consistent, averaging around 92.3 N. Moreover, it has been discovered that the underlying material displays an undulating surface (referred to as "A"), which is a consequence of the rolling process during sheet production. Additionally, the fracture morphology is primarily characterized by cleavage facets (referred to as "B") that are surrounded by regions where microvoids have merged together (referred to as "C").

The highest peel loads for different peel configurations on both sides (P12, P23, and P34). Additionally, the associated modes of fracture and surfaces of the fracture, respectively. According to P12, the peel tests performed in between the first and second layers show that the maximum peel loads are similar on both sides. Both examples have fractures close to the first layer's weld zone because of the layer's significant thinning and the weld zone's hardening, which raises localized stress in the area around the weld zone's margins. The fracture surfaces of P12 are on the retreating and advancing sides. Similar fracture surface features, such as a shoulder-SZ and a pin-SZ, are present on both sides. Cleavage facets dominate the fracture morphology in the pin-SZ.

The difference in peel test results between the advancing and retreating sides is more noticeable for P23 and P34 in comparison to P12. In the case of P23, a fracture occurs in the weld zone of the second layer on the advancing side, while the base material experiences cracks on the retreating side. These observations suggest that there is a more robust link being formed on the side that is moving away compared to the side that is moving forward. This observation supports the findings. The pin-SZ for P12 and the fracture surface of the second layer on the advancing side resemble each other. In particular, the main characteristic of the fracture morphology is the presence of cleavage facets. For P34, the advancing side sees fracture in the fourth layer's weld zone, while the retreating side sees removal of the whole fourth layer. The fracture surfaces of P34 are on the retreating and advancing sides. The pin-SZ and the fracture-containing base material are the two unique zones that make up the fracture morphology on the advancing side, according to the research. The base material fracture is a result of the spinning pin failing to penetrate the fourth layer. The major fracture morphology on the retreating side is divided into two separate areas: the pin-SZ (designated as "B") and the pin tip contact zone (designated as "A").

To summarize, the process of bonding in the multilayer connection devoid of defects is not uniform, leading to varying levels of bonding strength in each layer. Based on the analysis of the highest peel loads and types of fracture, it is observed that the third layer exhibits the strongest bonding strength on the advancing side due to the heat generated from the pin step (Q_4). However, because of the substantial material thinning, the initial layer has the poorest bonding strength. The P23 arrangement has the strongest bonding on the retreating side, whereas the P34 configuration has the weakest. Furthermore, compared to the foundation material, the weld zone is more brittle with cleavage facets in the weld zone.

4.5.7 Conclusions

- The μFSW was a successful method for joining thin aluminum alloy multilayers together. Surface-open tunnels, discontinuous semicircular hooks and surface flash are the three forms of surface defects that were detected, and it was determined that a narrow range of circumstances were necessary to establish perfect connections.
- The study found that optimizing the tensile load was significantly influenced by tool shape. By examining the major effect plots of four welding factors on the SN ratios, the ideal welding condition was found. These variables included plunge depth of 0.85 mm, welding speed of 3 mm/s, a rotation rate of 20,000 rpm, and the use of a stepped pin.
- The use of the optimal welding condition resulted in a multilayer joint without any defects. In this joint, the grains in the SZ (HAZ) were noticeably made smaller. From the top layer to the lowest layer, the length of bonding between neighboring layers grew shorter. Additionally, a more distinct TMAZ was seen on the side that was withdrawing compared to the advancing side.
- The defect-free joint's weld zone experienced substantial material hardening. While the first layer exhibited the most extensive hardened area, the microhardness measurements near the center of the weld were comparable across all layers.
- The adhesive strength of the defect-free junction with multiple layers varied among the four layers. This variation was influenced by the tool geometry, the location of the layers, and the welding position (whether it was the advancing or retreating side).[9]

4.6 MICRO-JOINING FOR BIOMEDICAL APPLICATIONS

Micro-joining in biomedical applications refers to the process of uniting small-scale components, often at the micro- or nanoscale, for different medical goals. This field is essential for the advancement of miniaturized medical devices, implants, sensors, and other biomedical instruments. Micro-joining in biological applications involves many important factors:

1. **Materials:**
   Biomedical micro-joining commonly entails the joining of several materials, including metals, ceramics, polymers, and composites. The selection of materials is based on the particular demands of the application, encompassing factors like as biocompatibility, mechanical characteristics, and usefulness.
2. **Joining Techniques:**
   (a) **Laser Welding:**
   Laser welding is frequently employed for micro-joining because of its accuracy, low heat input, and appropriateness for small-scale applications. The technology provides meticulous regulation of the welding procedure, facilitating the fusion of fragile or thermally vulnerable substances.

(b) **Micro-Resistance Spot Welding:**
This method is appropriate for connecting slender metal foils or wires in biomedical equipment. It provides rapid speed, little heat input, and excellent mechanical qualities.

(c) **Micro-Soldering:**
Micro-soldering is the process of connecting components using solder alloys on a very small scale. It is often employed in the assembly of small electrical devices and sensors for biological purposes.

(d) **Adhesive Bonding:**
Adhesive bonding is employed for the purpose of combining materials that are different from each other or when conventional techniques of joining are not practical. Biocompatible adhesives are necessary to guarantee compatibility with biological systems.

(e) **Micro-Friction Stir Welding:**
This solid-state joining approach is appropriate for the fusion of thermoplastics and other materials without inducing their melting. It provides benefits like reduced heat input, little distortion, and outstanding mechanical qualities.

(f) **Micro Ultrasonic Welding:**
It uses high-frequency vibrations to generate heat through friction at the interface of the joint, allowing for the fusion of thermoplastic materials. It is ideal for connecting tiny components with exceptional accuracy and efficiency.

3. **Biocompatibility:**
Biocompatibility is an essential factor to take into account in the field of biomedical micro-joining. For the purpose of preventing negative responses or rejection by the body's tissues, it is imperative that the components being joined together are both non-toxic and compatible with biological tissues. Materials and joining processes need to undergo thorough testing to guarantee they meet the requirements of biomedical standards and regulations.

4. **Sterilization Compatibility:**
Components utilized in medical devices or implants must be compatible with sterilization techniques such as autoclaving, gamma irradiation, or ethylene oxide treatment. To guarantee patient safety, it is crucial that joining procedures and materials retain their integrity and performance even after undergoing sterilization.

5. **Inspection and Testing for Quality:**
An essential component of biomedical micro-joining is quality control in order to guarantee the dependability and security of the connected components. Nondestructive testing techniques, including microscopy, ultrasonic testing, and X-ray inspection, are employed to assess the integrity of joints, correctness of dimensions, and qualities of materials.

Micro-joining is essential for the progress of biomedical technology, as it allows for the creation of small and implantable devices used in medical diagnostics, drug delivery, tissue engineering, and therapeutic applications.

4.6.1 Overview

Transmission laser joining shows significant promise in the fabrication and packaging of biomedical goods like sensors, microfluidic-biochips, and micro implants. Electronics and medical applications necessitate the micro-joining of materials that are different from one another. The significant issue of combining disparate materials necessitates the use of an appropriate joining method. When combining biomedical items, it is important to ensure that the joining procedure does not involve or produce any additional materials that are not compatible with living organisms. The laser joining procedure fulfills this criteria. Additionally, the laser joining technique is noncontact, hence preventing any contamination of the functioning sections of the biomedical goods. During this procedure, the laser energy may be carefully regulated inside the focus point, allowing for localized materials processing while minimizing the overall heat impact on the component. Minimizing heat impacted zones is crucial for biomedical goods. One benefit of laser processing is that it reduces tension in the materials that join due to its highly localized energy deposition. In addition, this technique of joining is very adaptable, has a reduced processing time, ensures a uniform level of quality, and may be repeated several times.

Prior research conducted by our team has demonstrated the effectiveness of using transmission laser micro-joining to join materials that are incompatible, such as polymers and metals or polymers and metal film–coated glass. The research used polyimide, titanium and glass since they have been demonstrated to be biocompatible materials and have the potential to be employed in enclosing tools for implanting brains. If the biological implants are placed in a biological environment, they must be packaged properly to avoid deterioration. Our lab has investigated the suitability of laser bonded micro-joints for encapsulating neural implants made of glass/polyimide and titanium/polyimide. Due to the highly encouraging results of previous studies, more research was conducted to enhance the strength of the joint. This section presented some of the outcomes of that research. The joining of dissimilar materials is contingent upon several diverse elements. The physical qualities exert a significant impact on the energy consumption for joining and heat transport. An investigation of the effects of mechanical stresses and material properties on the laser joint was carried out. An analysis of the causes of joint failure was conducted following mechanical testing in order to have a more comprehensive knowledge of the joints.

PVD is a process where atoms, ions, or occasionally molecules are condensed onto substrates. This condensation is achieved by activating them using high-energy electron, laser, plasma, and ion beams. The primary process control factors in PVD are the energy flow, background gas pressure, angle of incidence of the depositing particles (atoms and/or ions), and substrate temperature. The deposition method and process parameters have a significant influence on the microstructure and characteristics of films. The EB-PVD technique utilizes high-energy electron beams produced by electron guns to evaporate and melt the components of the coating. The substrate inside the vacuum chamber is likewise warmed by these beams, which helps to improve the adherence of the coating. This technique exhibits features such as comparatively rapid deposition rates, precise control over composition, and little contamination. The technique often yields coatings with a high-quality surface finish and a consistent microstructure. The coatings provided by CA-PVD exhibit an extensive range of morphological and microstructural features. The CA-PVD technique allows for the production of a significantly greater level of ionization compared to conventional ion-plating processes, resulting in improved adhesion of film.

Variations in the film preparation can lead to system interfaces with different laser joint strengths. The thin-film thermal conductivity is influenced by microstructural characteristics. Research already conducted indicates that EB-PVD thin films have incredibly low thermal conductivity, which varies depending on the coating thickness. Moreover, various coating techniques provide films of varying surface roughness. The adhesion strength is influenced by the alteration in surface area caused by substrate roughness. Increasing the surface area will result in a greater number of bonds developed in the space between the substrates. Rather than the formation of strong connections, the number of bonds is responsible for the different adhesion properties of metal and PI. Preliminary research indicates that vapor-deposited films often exhibit internal stresses, known as residual stresses, which frequently cause the substrates to spontaneously delaminate and shatter. This sort of coating with inadequate adhesion will have a negative impact on the laser joining process.

4.6.2 Experimental Methods

A thin titanium layer with a thickness of about 200 nm was applied over glass in order to aid the bonding process with polyimide. It was a coupling agent made of metal film. The methods employed for film deposition were cathodic arc physical vapor deposition (CA-PVD) and electron beam evaporation (EB-PVD). A 0.2 mm thick PI (Imidex) film from Westlake Plastics was combined with a Pyrex Borosilicate 7740 glass wafer with a thin layer of Ti film covering it (0.5 mm thick) from Corning. The process of depositing a titanium (Ti) coating on glass using electron beam evaporation (EB-PVD) was carried out using a Kurt Lesker evaporation system. The rate of deposition was around 1.5 angstroms per second (Å/s). A base pressure of 1×10^{-3} Pa was maintained while the substrate was positioned at room temperature. During the process of CA-PVD, a layer of titanium (Ti) was applied onto a glass substrate using the metaplas ionon system (MZR 303U) at a rate of deposition of 5–10 angstroms per second (Å/s). The pressure at the initial stage was approximately 1×10^{-3} Pa. The coatings were produced at the East Lansing, Michigan, USA-based Fraunhofer Center for Coatings and Laser Applications. A 1100 nm-wavelength continuous wave (cw) Yb-doped fiber laser beam was used for the joining operation. The JDS Uniphase Inc.-manufactured IFL-25 fiber laser system was used. This system had a continuous wave output power of 25 W and produced a beam that was diffraction restricted. The 200 μm-diameter laser beam showed a Gaussian power distribution. A gearbox connecting design was utilized, wherein the energy of the laser pierced the transparent component and was taken up by the absorbent substance, resulting in direct heat induction at the interface.

Two 11 mm thick borofloat glass blocks (from Howard Glass Company) were used for the sample processing in order to secure the sample. To form the connections, a clamping pressure of 415 kPa was used. Consistent application of pressure was crucial for achieving a high-quality joint. Prior to joining, the materials underwent ultrasonic cleaning using iso-propanol and acetone, followed by drying with nitrogen gas. The rough side of the polyimide sheets was facing the beam when they were placed. Initially, the laser beam was directed toward the titanium-coated glass wafer. Before entering the PI film and starting the joining process at the interface of the joining materials, the beam passed through a borofloat glass plate. After directing the beam toward the surface, it was intentionally adjusted to be out of focus by a specific distance, often ranging from 1.5 to 2.0 mm. The beam was defocused to mitigate the risk of overheating, degradation, and burning of the polymers by lowering the laser spot's central intensity. By varying the laser power and speed but maintaining consistent defocusing, the ideal joint strength was found. The selection of laser settings is discussed by researchers in who demonstrated that optimal bonds may be established across a wide range of speeds with a consistent power level. Additionally, bonding at greater power levels necessitates higher speeds. In their study, Some researchers investigated how laser bond strengths were affected by laser power and speed. They discovered that lower speed and lower laser power resulted in stronger bond strength. During the process of CA-PVD, a joint formed as a result of 10% power fluctuations occurring at a rate of 100 mm/min. However, this variation caused a decrease in strength to 50%. EB-PVD has a significantly narrower variety of processing factors for achieving effective bonding compared to CA-PVD. A slight deviation in power resulted in a much weaker or nonexistent connection. During the joint-making process, it was necessary to apply heat to the materials at the starting point for a duration of 0.05 seconds in order to achieve a consistent joint line's width. A laser joint line measuring 6.5 mm in length and 0.3 mm in joint width was created by using a laser with a 1.0 W power output and 100 mm/min scanning speed. The breadth of the joint was determined using an optical microscope following the preparation of the sample. A laser line was used to create laser connections between the 30 mm 10 mm chunks of material in the overlapping zone.

A submicron tester with six axes was used to measure the samples' joint strength. The micro-mechanical/thermal testing device is designed for general-purpose use. With the system, the tested specimen may be subjected to displacements or weights in six directions—three orthogonal translations and three rotations. The device is entirely controlled by a computer. Additionally, load-displacement data is gathered by the machine's software from the load cell and capacitance gauge that are attached to it. The height of the micro-tester grips was changed to preserve a straight joint line and decrease the bending moment. Furthermore, the polyimide undergoes deformation when it is installed in the micro-tester, resulting in the application of compression force on the specimen owing to the clamps. By closely monitoring the forces and moments in all six axes and adjusting the location of the stages either before to or during sample mounting, all extraneous mounting forces were eliminated, resulting in a complete elimination of undesired forces and moments. The samples underwent uniaxial stress by turning on the motor that moves along the length of the sample's direction. Testing was done on the samples at a speed of 0.01 mm/s.

This work utilized Atomic Force Microscopy (AFM) to analyze the quality and roughness of the substrate surface. The Scotch tape test was used to determine how well the metal film adhered to the substrate. SEM and optical microscopy were employed to analyze the failure mechanisms of the laser joint, and these findings were then connected with the laser joint strength. The tool used for the AFM examination was the Digital Instruments Nanoscope III. The roughness investigation was done using a profilometer (Dektek 3030) because of the CA-PVD surfaces' irregularity. After a comprehensive optical microscope inspection of the failed joint surfaces, a SEM was employed to gather more information regarding the processes behind the laser joints' failure. The materials were imaged using a Hitachi S-2400 electron microscope to acquire SEM pictures. The device utilizes a tungsten filament operating at a peak voltage of 25 kilovolts. The failing laser joint surfaces were analyzed using an Energy-Dispersive Spectrum (EDS) detector. In order to prevent charging effects, prior to performing SEM and EDS, glass surfaces coated with PI and Ti were sputter-coated with a thin layer of gold (Au) using an Ernest F. Fullam sputter coater.

4.6.3 Outcomes

4.6.3.1 Surface Examination of Ti-Coated Glasses

Coatings generated by EB-PVD had a highly polished surface quality, while coatings generated by CA-PVD displayed rough surfaces with the presence of flaws in the form of macro-particles. One frequent feature of the CA-PVD process is the presence of macro-particles on the surface. Using AFM images with a size of (10 × 10) μm^2, which were produced by the EB-PVD approach, the coated surfaces' root mean square roughness (RMS) was determined to be 1.5 nm. To take into consideration the unevenness of the CA-PVD surfaces, a profilometer with a larger scanning area was used for the roughness investigation. The surface has an average roughness (R_a) of 125 nm, the analysis found.

4.6.3.2 Measuring Bond Strength

The shear strength of the laser joints created by employing polyimide and thin Ti film–coated glass was (3.7 ± 1.7) MPa in the EB-PVD method and (14.0 ± 2.5) MPa in the CA-PVD approach. The coating processes exert a substantial influence on both the development and joint strength. The cohesive force of the joint is attributed to the creation of chemical bonds at the interfaces between the metal and polymer. Prior research undertaken by our team on comparable connecting systems such as Ti and Ti/PI coated (CA-PVD) glass/PI shows that strong chemical bonds of the Ti-O and Ti-C kinds are formed at the laser-jointed surfaces.

4.6.3.3 Failure Analysis

Flashes were detected throughout the joint lines when EB-PVD samples were joined by laser, showing that there is inadequate heat transfer between the materials and a lot of heat production in the joint area. The composite picture displays clear signs that the PI is damaged, shown as a configuration resembling a network formed along the joint line.

Following testing for lap shear, fragments of shattered glass were discovered on the EB-PVD joints technique. This suggests that the strength of the glass has decreased as a result of the tension caused during the construction of the joint. The SEM images of titanium-coated glass surfaces subjected to lap shear testing revealed microfractures and porous polyimide (PI) residue on the joints. Through the glass, the fissures could also be seen on the other side of the undamaged joint. Upon conducting lap shear testing, it was observed that the PI side exhibited a presence of material layers on some joints, accompanied by the presence of fractures. The EDS examination of the adhering substance conclusively verifies that it consisted of a thin coating of glass.

After undergoing lap shear testing, the joint's surfaces that experienced failure in the CA-PVD technique were examined using an optical microscope. The optical microscope images of a sample that underwent testing for lap shear. Broken shards of glass were detected along the joint line. Using a SEM, the joint lines of the specimens used for lap shear testing were inspected. The failing glass surfaces exhibited polyimide residues in the exposed joint locations.

4.6.4 Discussions

The properties of the layer on the glass substrate were seen to have an impact on the laser bonding strength. The particles placed in the CA-PVD exhibited greater energy levels compared to those in EB-PVD due to the increased rate of deposition in CA-PVD. In CA-PVD, the deposition rate ranged from 5 to 10 angstroms per second, while in EB-PVD it was 1.5 angstroms per second. Consequently, the particles of CA-PVD showed a higher sticking coefficient, resulting in improved attachment to the substrate made of glass. There was an increase in surface roughness in the CA-PVD film. Better mechanical interlocking, a larger contact area at the interface, and more chemical bonds were produced by increased roughness, which improved laser joint strength.

The adherence of the EB-PVD film over the glass substrate was unsatisfactory, and the film exhibited insulating properties. The metal film generated by electron beam physical vapor deposition (EB-PVD) exhibited a highly polished surface, as confirmed by the AFM study. The film exhibited low heat conductivity, resulting in the formation of hotspots caused by localized film decohesion. Intense heat was produced during the joining process, resulting in bright flashes along the lines where the materials were joined.

Glass failure was seen during the lap shear test in both ways, resulting from the residual stress caused during the bonding process on the glass. Thermal residual stress occurred in the joint region during the joining process because there were differences in the coefficients of thermal expansion (CTE) between the polyimide, glass, and Ti covering. When compared to glass, titanium has a greater CTE, which means that as it cools, it will exert a pulling force on the glass, which is already vulnerable to tension. Glass splits when the transitory residual stress surpasses the tensile strength of the glass. The EB-PVD approach resulted in increased stress on the glass, which can be attributed to the presence of microcracks.

4.6.5 Conclusions

Thin titanium films were developed on glass wafers using EB-PVD and CA-PVD methods to get the films ready for connecting with polyimide sheets using a transmission laser. The CA-PVD method yielded titanium layer at a rate of 5–10 angstroms per second, whereas in the EB-PVD process it was 1.5 angstroms per second. The adherence of the titanium film on glass achieved using the CA-PVD technique surpassed that of the EB-PVD approach. The reason for this is probably the result of a faster rate of film deposition, resulting in the CA-PVD particles having a higher adhesion or sticking coefficient to the substrate than the EB-PVD film. The laser joints created by CA-PVD were about four times stronger than those produced by EB-PVD. The Ti film formed by EB-PVD exhibited inadequate laser joining qualities due to very smooth surface characteristics, limited initial adhesion to the substrate, and a propensity to produce hotspots as a result of localized film decohesion. However, the metal film created by CA-PVD was more successful due to its robustness, greater adhesion, and rougher substrate, which positively influenced the joint strength.[10]

REFERENCES

1. Tsai, J.-T., Chang, C.-C., Chen, W.-P., and Chou, J.-H., 2016, "Optimal parameter design for IC wire bonding process by using Fuzzy Logic and Taguchi method," *IEEE Access*, vol. 4, pp. 3034–3045, doi:10.1109/ACCESS.2016.2581258
2. Charles, H. K., 2001, "Microelectronic packaging: Electrical interconnections," in *Encyclopedia of Materials: Science and Technology*, Elsevier, pp. 5616–5635. doi:10.1016/B0-08-043152-6/00981-5
3. Charles, H. K., 2009, "Advanced wire bonding technology: Materials, methods, and testing," in *Materials for Advanced Packaging*, Springer US, pp. 113–179. doi:10.1007/978-0-387-78219-5_4
4. Wang, J., Wang, Q., Wu, Z., Wang, D., and Cai, J., 2017, "Solid-state-diffusion bonding for wafer-level fine-pitch Cu/Sn/Cu interconnect in 3D integration," *IEEE Transactions on Components, Packaging and Manufacturing Technology*, vol. 7, no. 1, pp. 19–26, doi:10.1109/TCPMT.2016.2626794
5. Yan, S., Li, Z., Song, L., Zhang, Y., and Wei, S., 2023, "Research and development status of laser micro-welding of aluminum-copper dissimilar metals: A review," *Optics and Lasers in Engineering*, vol. 161, p. 107312, doi:10.1016/j.optlaseng.2022.107312
6. He, X., Elmer, J. W., and DebRoy, T., 2005, "Heat transfer and fluid flow in laser microwelding," *Journal of Applied Physics*, vol. 97, no. 8, doi:10.1063/1.1873032
7. Klages, K., Ruettimann, C., and Olowinsky, A., 2003, "Laser beam micro welding of dissimilar metals," in *International Congress on Applications of Lasers & Electro-Optics*, Laser Institute of America, p. M202. doi:10.2351/1.5060103
8. Sen, M., Shankar, S., and Chattopadhyaya, S., 2020, "Micro-friction stir welding (μFSW): A review," *Materials Today: Proceedings*, vol. 27, pp. 2469–2473, doi:10.1016/j.matpr.2019.09.220
9. Wang, K., Khan, H. A., Li, Z., Lyu, S., and Li, J., 2018, "Micro friction stir welding of multilayer aluminum alloy sheets," *Journal of Materials Processing Technology*, vol. 260, pp. 137–145, doi:10.1016/j.jmatprotec.2018.05.029
10. Sultana, T. et al., 2009, "Study of two different thin film coating methods in transmission laser micro-joining of thin Ti-film coated glass and polyimide for biomedical applications," *Journal of the Mechanical Behavior of Biomedical Materials*, vol. 2, no. 3, pp. 237–242, doi:10.1016/j.jmbbm.2008.07.004

5 Challenges and Opportunities in Micro-Joining of Materials

5.1 INTRODUCTION

With the growing need for high-performance, multifunctional, and small components in industries including electronics, aerospace, medical devices, and automotive systems, micro-joining is essential to modern manufacturing. It entails the exact connecting of parts that are usually between a few millimeters and micrometers in size, frequently incorporating different materials and necessitating a high dependability and precision level.

Miniaturization is one of the main factors influencing the use of micro-joining methods. Traditional connecting techniques are becoming inefficient or unfeasible as electronic devices and systems get smaller. Compact and integrated systems like microelectromechanical systems (MEMS) and high-density electronic packages may be produced more easily because of micro-joining, which makes it possible to assemble components reliably at micro-scales.[1]

The ability to link different materials is a major benefit of micro-joining as well. This is crucial for modern devices that mix semiconductors, metals, polymers, and ceramics to create customized features.[2] Sensitive components may be harmed by the considerable mechanical or thermal stress that conventional connecting procedures may cause. Localized, low-heat procedures that preserve the integrity of the base materials are provided by micro-joining techniques such as diffusion bonding, laser welding, and ultrasonic bonding.[3]

Important high-tech businesses also benefit from micro-joining. For instance, wire bonding and flip-chip bonding are essential to printed circuit board (PCB) construction and semiconductor packaging in electronics.[4] In the medical industry, where joint dependability, biocompatibility, and cleanliness are critical, micro-joining methods are crucial for creating tiny implants, sensors, and surgical tools.[5] Micro-joining guarantees the high-precision assembly of micro-actuators and sensors needed for navigation and control systems in the aerospace and defense industries.[6]

Additionally, micro-joining facilitates automation and high-throughput production, which is consistent with industry 4.0 objectives. Robotic systems can use processes like resistance micro-welding and laser micro-welding to increase efficiency and consistency.[7] By lowering energy input, cutting waste, and creating cleaner processing environments, these methods also support material efficiency and sustainability.[8]

DOI: 10.1201/9781003502395-5

5.2 MATERIALS USED IN MICRO-JOINING

5.2.1 Metals

Because of their great mechanical strength, ductility, and electrical and thermal conductivity, metals are the most often connected materials in micro-joining. Copper, aluminum, gold, titanium, and stainless steel are common metallic materials used in biomedical implants, automobile sensors, and microelectronics.

(a) **Micro-Joining Methods**:
Resistance micro-welding, laser micro-welding, ultrasonic bonding, and micro-TIG welding are commonly used for metals.
(b) **Applications**:
Wire bonding in semiconductor packaging (e.g., Al-Au, Cu-Cu), battery connections, MEMS actuators, and biomedical implants.

The oxidation of reactive metals and the requirement for exact heat management to prevent melting of tiny components present a problem. According to research, ultrasonic welding works very well for combining thin metallic foils and interfaces between incompatible metals.[9]

5.2.2 Polymers

Micro-devices are using polymers more and more because of their flexibility, low cost, lightweight nature, and resilience to chemicals. However, connecting becomes more difficult because of their low surface energy and limited thermal stability.

(a) **Micro-Joining Methods**:
Laser welding (especially transmission laser welding), ultrasonic welding, and adhesive bonding are commonly used.
(b) **Applications**:
Microfluidic devices, flexible electronics, biomedical sensors, and polymer-based optical components.

To increase joint strength, polymers frequently need to have their surfaces modified or intermediate layers added. For microfluidics, laser transmission welding has demonstrated efficacy in precisely attaching transparent polymers such as PC and PMMA.[10]

5.2.3 Composites

Because of their specialized mechanical and thermal qualities, composites—in particular, fiber-reinforced polymers (FRPs) and metal matrix composites (MMCs)—are preferred in the biomedical, electronics, and aerospace industries. However, creating reliable and robust micro joints is difficult due to their heterogeneous and anisotropic character.

(a) **Micro-joining Methods**:
Ultrasonic welding, micro-resistance welding, and laser micro-joining are applied, often with surface conditioning.
(b) **Applications**:
Sensors, structural miniaturized components, and smart implants.

To prevent fiber damage, matrix breakdown, and resin delamination, more caution is required. Localized heating using laser-based methods is one recent development that enables selective joining with little damage.[11]

5.2.4 Ceramics

Because of their exceptional hardness, resilience to high temperatures, and biocompatibility, ceramics are used in MEMS devices, high-temperature sensors, and medical implants. Their lack of plastic deformation, low-heat conductivity, and brittleness, however, make it challenging to combine them using traditional techniques.

(a) **Micro-Joining Methods**:
Transient liquid phase bonding, diffusion bonding, brazing with active fillers (e.g., Ti), and laser joining with glass interlayers.
(b) **Applications**:
Dental and orthopedic implants, sensor housings, and electronic packaging.

For instance, to increase wetting during ceramic-to-metal bonding and address thermal expansion mismatches, glass or metal interlayers are frequently used.[12]

5.3 FUNDAMENTALS OF MICRO-JOINING

The term "micro-joining" describes a group of joining techniques used to put together features or components that fall within the micro and millimeter range; these are usually smaller than 1 mm or have joint dimensions at the micron level.[9] These methods, which are intended for use with thin films, tiny parts, or delicate wires, frequently call for high accuracy, localized energy input, and little mechanical or thermal distortion. They usually range in size from 1 μm to around 1 mm. In precision engineering, micro-joining is essential because of the joint's tiny size and the materials' sensitivity, which necessitate rigorous control over alignment, pressure, and heat input. Particularly in microelectronics and biological applications, mistakes in micro-joining can lead to the complete system failing.

The term "conventional joining" describes the use of conventional methods to join parts, usually on a millimeter to meter scale. In contrast to micro-joining, these procedures are usually more energy-intensive and need less precise tolerances because they are intended for bigger components. Their range of dimensions is more than 1 mm to meters. A wide variety of materials are employed here, such as ceramics, polymers, and structural metals. Arc welding (MIG, TIG), brazing, soldering, mechanical fastening, and adhesive bonding are the methods employed here.

TABLE 5.1
Micro-Joining vs. Conventional Joining—A Comparison

Feature	Micro-Joining	Conventional Joining
Size of components	< 1 mm (micron to sub-mm scale)	> 1 mm (mm to meters)
Precision	Very high	Moderate to high
Energy input	Localized, low	Higher, less localized
Applications	Electronics, MEMS, biomedical	Automotive, aerospace, construction
Heat-affected zone	Very small or minimal	Larger
Joining techniques	Laser micro-welding, ultrasonic, etc.	Arc welding, brazing, adhesives, etc.
Typical materials	Thin metals, ceramics, polymers	Structural metals, bulk materials

Common uses include pipelines, building parts, automobile frames, and aviation structures. A wide variety of materials are employed here, such as ceramics, polymers, and structural metals. Arc welding (MIG, TIG), brazing, soldering, mechanical fastening, and adhesive bonding are the methods employed here. Construction components, pipelines, automobile frames, and airplane constructions are examples of typical uses (Table 5.1).[13]

5.4 CLASSIFICATION OF MICRO-JOINING PROCESSES

The underlying joining mechanism is used to categorize micro-joining approaches. Fusion joining, solid-state joining, mechanical joining, and hybrid joining are the four main categories. Certain materials, joint geometries, and application situations are best suited for each class.

5.4.1 Fusion Micro-Joining

Melting the base materials—with or without filler—to create a junction when they solidify is known as fusion joining. This high-temperature procedure frequently calls either shielding gas or a controlled environment and may entail a heat-affected zone (HAZ). This method works well with thermoplastics and metals. Among the methods used in fusion micro-joining are **electron beam micro-welding** (which provides deep penetration and a very narrow HAZ), **laser micro-welding** (which uses a focused laser to melt material at the joint line with micron precision), and **micro-soldering** (which involves melting a filler (such as Sn-Pb, Au-Sn) without melting base metals). Microelectronics, optical assemblies, and medical devices are examples of typical uses.[14]

5.4.2 Solid-State Micro-Joining

The foundation materials are not melted during solid-state joining. Rather, it depends on frictional heat, plastic deformation, or diffusion. It is a technique that produces excellent junction quality for different materials, needs no melting, and has minimum thermal distortion, making it perfect for delicate and small components. **Thermocompression bonding** (heat and pressure cause atomic interdiffusion), **ultrasonic bonding** (high-frequency vibrations cause interfacial friction and bonding at low temperatures), and **diffusion bonding** (performed at high temperature under pressure for extended time) are some of the methods used in solid-state micro-joining. Common uses include hermetic sealing, MEMS packaging, and wire bonding (Au-Au, Al-Al).[9]

5.4.3 Mechanical Micro-Joining

In order to produce joints without melting or atomic-level bonding, mechanical joining uses interlocking, deformation, or friction-based processes. It is a technique that doesn't require any heat input, is frequently reversible, and employs simpler setups, but its miniaturization is limited. **Micro-riveting** (deforming micro fasteners to keep pieces together), **micro-screws** and **clips** (used in tiny assemblies where disassembly may be required), and **micro-stapling** (used in electronics or healthcare devices) are a few of the mechanical micro-joining techniques. Micromechanical systems and tiny consumer electronics are examples of typical uses.[13]

5.4.4 Hybrid Micro-Joining

Hybrid joining maximizes joint performance by combining two or more joining methods (such as fusion and diffusion or thermal and mechanical joining), particularly in complicated or different material systems. It is a technique that improves joint strength and dependability, gets around the drawbacks of conventional approaches, and works well with joints that are dissimilar or have several layers. Among the methods utilized in hybrid micro-joining are **adhesive + mechanical** (a hybrid bonding method for flexible electronics), **ultrasonic-assisted soldering** (which enhances wetting and minimizes voids), and **laser-assisted diffusion bonding** (which combines localized heating and pressure). Typical uses include implantable devices and heterogeneous integration in microelectronics.[2]

5.5 MICROSTRUCTURAL IMPLICATIONS OF MICRO-JOINING PROCESSES

5.5.1 Grain Structure Alteration

The grain structure may alter significantly as a result of localized heating during micro-joining: Because of recrystallization and grain development brought on by heat exposure, grain coarsening frequently happens in the fusion zone. On the other hand, fine-grained or even amorphous structures may result from the quick solidification of some micro-welding processes (such as laser micro-welding).[15]

5.5.2 Formation of Intermetallic Compounds (IMCs)

In the context of dissimilar metal micro-joining (e.g., Cu-Al, Ni-Ti), the formation of intermetallic phases like Cu_6Sn_5 or FeAl at the interface can frequently result in brittleness.[16] The detrimental effects of excessive IMC layer growth are influenced by time-temperature profiles.

5.5.3 Residual Stresses and Microcracks

Elevated thermal gradients in micro-joining can lead to residual tensile stresses, which may encourage the initiation and propagation of microcracks, particularly in brittle materials.[17]

5.6 HEAT-AFFECTED ZONES OF MICRO-JOINING PROCESSES

Despite the lower energy inputs associated with micro-joining processes, they still produce a HAZ, which is a thermally altered area in the base material next to the fusion zone.

5.6.1 HAZ Width and Characteristics

The width of the HAZ in micro-joining typically measures in the micrometer range, particularly in processes such as laser or resistance welding.[18]

Even though it is small, the HAZ can undergo phase transformations, grain growth, or precipitate coarsening, influenced by the material system.

5.6.2 Material-Specific HAZ Effects

(a) **Aluminum Alloys**:
Alloys that undergo precipitation hardening may experience overaging in the HAZ, leading to a decrease in strength.[19]

(b) **Titanium Alloys**:
The occurrence of brittle alpha-case and grain coarsening is frequently observed in the HAZ.

(c) **Stainless Steel**:
The potential for sensitization resulting in intergranular corrosion in the HAZ if not cooled promptly.

5.6.3 Control Strategies

A variety of strategies exist to alleviate negative microstructural alterations and HAZ impacts in micro-joining.

(a) Utilization of ultrafast laser pulses or electron beams to precisely target heat input and reduce the HAZ.
(b) Utilization of inert gas shielding or vacuum environments to inhibit oxidation and avoid undesired phase development.
(c) Implementation of post-joining heat treatment, such as stress relief annealing, aimed at minimizing residual stresses.

5.7 TECHNICAL CHALLENGES IN MICRO-JOINING

5.7.1 Material Compatibility and Dissimilar Material Joining

The challenge of joining dissimilar materials in micro-joining remains both persistent and complex, particularly as it becomes more prevalent in advanced microelectronic packaging, MEMS, and biomedical device applications. Micro-scale devices frequently comprise a variety of materials that exhibit significantly different physical, thermal, and chemical characteristics, such as metals, ceramics, polymers, and semiconductors. For example, a standard sensor might necessitate the connection of silicon to stainless steel or copper, while a bio-implant could entail the adhesion of titanium to a biocompatible polymer. The variations in coefficients of thermal expansion, surface energies, melting temperatures, and chemical reactivity pose considerable challenges in these combinations. A discrepancy in coefficient of thermal expansion can result in the buildup of significant residual stresses throughout thermal cycling, particularly during processing or in operational conditions. For instance, silicon exhibits a coefficient of thermal expansion of roughly 2.6×10^{-6}/K, whereas copper has a value around 17×10^{-6}/K. This difference can lead to mechanical stresses that may result in cracking or delamination at the joint interface during the cooling phase following bonding.

Beyond thermal challenges, chemical incompatibilities at the interface can lead to the creation of brittle IMCs or diffusion layers, which undermine the mechanical integrity and electrical performance of the joint. In gold-aluminum bonding systems, which are frequently used in wire bonding, the creation of intermetallics like Au_5Al_2, often referred to as "purple plague," is recognized as a significant factor leading to catastrophic failure due to its inherent brittleness. Establishing wetting between different materials presents significant challenges; metals can rapidly oxidize in the presence of air, while ceramics generally show limited wetting with molten metals due to their ionic or covalent bonding properties. To address these challenges, it is common to utilize interlayers (like Ti, Cr, or Ni) or to create functionally graded materials that facilitate a gradual transition between varying material properties. Nonetheless, these solutions introduce additional layers of complexity and elevate the number of interfaces, each of which may be vulnerable to failure. Moreover, the creation of these intermediate layers necessitates extra process steps such as sputtering, electroplating, or diffusion bonding, each of which must be meticulously regulated at the micro-scale.

It is crucial to recognize that at micro or nano scales, surface effects take precedence over bulk material properties, thereby complicating predictions and the development of processes. Surface contamination, the adsorption of ambient gases, and local oxidation significantly hinder the formation of strong, reliable bonds between dissimilar materials. Traditional bonding theories grounded in bulk thermodynamics frequently do not adequately address the complexities of atomic-scale interactions at these dimensions. Consequently, materials engineers are compelled to investigate innovative modeling techniques—like molecular dynamics simulations and first-principles calculations—to anticipate interfacial behavior. Even with progress made, the effective integration of different materials at micro-scales continues to pose a considerable technical hurdle that restricts the scalability, performance, and durability of micro-devices.[20–22]

5.7.2 Heat Input and Thermal Management in Micro-Joining

Managing thermal conditions in micro-joining is a critical and technologically difficult aspect of the process, especially for parts with small dimensions and high thermal sensitivity. Most micro-joining methods (laser welding, thermosonic bonding, resistance welding, and reflow soldering) require precise application of heat to achieve bonding. While the HAZ can rapidly spread in macro-welding due to heat conduction, making these processes less precise, in micro-joining processes, the situation is completely the opposite, and a slight excess of thermal energy can cause highly localized damage. This is due to the much lower thermal mass of said materials, and hence their inability to deal with heat better. Parts such as thin films, wires, and MEMS structures might deform, melt, or even undergo a change in its phase if they are subjected to unregulated thermal gradients. For systems containing polymers or organic coatings, elevated temperature can induce outgassing, burning, or delamination. This is of particular concern in hybrid microsystems, where organic and inorganic components are placed in close proximity.

Further, the presence of different materials complicates thermal issues due to disparate thermal conductivities. For example, when joining a readily conductive metal, e.g., copper (≈400 W/m·K), to a polymer, such as polyimide (≈0.2 W/m·K), the heat management has to be managed carefully such that the polymer does not become too hot while providing sufficient heat to bond to the metal surface. In micro-joining, thermal diffusion lengths are often only a few μm square, thereby making it extremely easy to heat up regions that one did not intend, possibly destroying side-by-side components—some of which may be very delicate microelectronics, or a micro encapsulated system. Moreover, thermal effects can also affect the interface; increased heating or prolonged heating can induce grain growth, recrystallization, and intermetallic compound formation, all of which may adversely affect joint strength or reliability. In applications such as in implantable medical devices or aerospace systems where failure cannot result, thermal uniformity and containment is become a key requirement.

From the additive process control point of view, the real-time in situ thermal monitoring during the micro-joining process is a major challenge. Conventional thermocouples and infrared sensors often are too large and/or not sufficiently accurate for micro-scale applications. The join process in many cases occurs without being directly monitored, relying exclusively on previously set process parameters. Therefore, high-resolution thermal sensors, thermoreflectance imaging, or pyrometric systems are increasingly needed in micro-joining systems. Generally, finite element analysis (FEA) simulations are used to describe thermal profiles; however, such models often have padlocks due to assumptions for material, properties of materials, boundary conditions when scaling down to micro-levels. This possibility is aggravated in an industrial environment, where no time is left for profile optimization or set up-trials if high-throughput conditions are requested. Pulse-based methods for energy delivery (pulse Nd:YAG, femtosecond laser) are presently used to realize the localization of the heating with a high precision in space and time. The use of these types of systems can also help to restrict the HAZ and reduce unwanted thermal damage, although it is more expensive to implement.

An important issue is effect of repetitional thermal cycling on the reliability of the micro joint. Many other devices stay in service at widely varying temperatures for long periods of time (e.g., from 25°C to 125°C in automotive electronics), leading to thermal fatigue, crack propagation, or solder creep. Therefore, the problem of thermal management includes not only the joining but also the behavior of the joint when it is subject to different thermal loading during its service life. To prevent these thermal hazards, a holistic thermal risk reduction program must be developed, which also includes proper selection of materials, joint design and process optimization. So, the successful micro-joining demands not only a balance of getting a sufficient inputting heat for bond formation, but also that the function or structure of the whole system keeps intact.[23–26]

5.7.3 Size Effects and Precision Requirements in Micro-Joining

A major problem in micro-joining is caused by the strong influence of size effects when a process is scaled down from macro to micro- or even nanoscale. At this scale, material properties have behaviors that are vastly different to those of traditional manufacturing, making established joining methods impractical or ineffective. At larger scale, gravity and inertia become negligible and surface forces, namely, van der Waals forces, surface tension, and capillary effects predominate. These forces can interfere with the intended dynamics of bonding or welding operations. In micro-soldering, the viscous forces of the molten solder leads to nonwetting, in particular if the substrate dimension is in the range of a few micron due to high substrate subtractive. Moreover, scaling laws predict that surface area-to-volume ratios will increase markedly, and as a consequence, heat losses, oxidation kinetics and susceptibility to surface contamination increase, which complicates the joining process and reduces the reliability of joints.

An important aspect of size is that there is a requirement for very accurate alignment and location of parts. For micro-joining, even a deviation in the sub-micrometer range is not acceptable, as required for example in optical MEMS, biological implants, or microfluidic applications, where exact operation alignment is essential. In micro-optics packaging, 2–5 μm of misalignment will cause the big signal loss or image distortion. Traditional tooling methods and positioning systems have difficulty achieving, and sustaining, this level of precision in reliable and unambiguous manner, especially for mass production situations. This involves the use of nano-positioning stages, vision-guided robotic systems, and actuators with a feedback control, all at a high cost and requiring expertise in operation. Additionally, maintaining accuracy during bonding, especially in the presence of thermal expansion or machine vibration, introduces additional complications. For example, the alignment achieved during a setup process can be misaligned due to movement, thermal distortions or tooling wear during bonding, particularly in methods like resistance welding or thermosonic bonding.

In addition, the form of the elements to be joined also influences heat generation, the application of pressure and flow of the material. Local heating, mechanical stress-induced breakage or poor alignment of such less than a few microns thick elements like thin-film layers or nanowires, are realistic hazards. Due to the nonuniformity of force over these small interfaces, debonding or void are likely to occur, thus degrading both the electrical and mechanical performance. It is a challenge to produce instruments that can provide homogeneous pressure or energy to the small contact areas. The common tools are either not accurate or too much sturdy for such tasks. High-level, MEMS-based tooling or compliant microgrippers have been designed to overcome these challenges but these are expensive and not easily scalable.

These get more complicated by the difficulties of in-line characterization on micro and nano scales. Unlike macro joining, where the weld can be readily examined visually or using conventional nondestructive evaluation (NDE) techniques such as ultrasonic or radiography, micro-joining often lacks available and reliable in situ monitoring methods. Minuscule defects, such as microcracks, voids, and lack of fusion, remain difficult to detect using nondestructive methods without resorting to scanning electron microscopy (SEM), focused ion beam (FIB) cross-sectioning or acoustical microscopy. However, these methods are extremely time-consuming, expensive and not suitable for achieving in-process total inspection. For this reason, process engineers must use predictive models and process controls that rely on another variable such as electrical resistance, thermal feedback or acoustic emission during the joining process. However, indirect indicators may not predict such small scales reliably, so there may be a mismatch in process validation and actual joint performance.

In addition, the effects of microstructural gradients are exacerbated at smaller length scales. In bulk, properties can be averaged over large volumes, but at a microscale, small microstructural fluctuations—such as grain boundaries, phase distributions, or inclusions—can have a strong impact on mechanical properties. For instance, the grain size at the joint interface can be largely changed in the micro-joining process, especially in the situation of ultrafast solidification or cooling. Microstructural modifications can make the material brittle or weak in some areas, which shall not be observed until its usage under mechanical or thermal efforts. Therefore, it is essential to understand the materials science that occurs at micro and nano scales to optimize the joining process.[27–30]

5.7.4 Surface Cleanliness and Preparation in Micro-Joining

The cleanliness of surfaces plays a crucial role in determining the effectiveness of micro-joining processes, especially due to the significant surface-to-volume ratio inherent in components at the micro- and nanoscale. In micro-joining, the interface between two materials encompasses merely a few square microns of contact area, and any contamination or oxide formation on that surface can significantly impact bond integrity. Contaminants measuring just a few nanometers—like native oxides, adsorbed moisture, hydrocarbon films, or processing residues—can serve as obstacles to diffusion, hinder wetting, or trigger delamination following the joining process. This challenge is particularly pronounced in joining methods such as diffusion bonding, thermocompression bonding, and ultrasonic welding, where close atomic contact or plastic deformation is essential for effective joint creation. Even minimal contamination can hinder the establishment of metallic or covalent bonds, thereby diminishing the mechanical strength and conductivity of the joint. Metal oxides, which develop quickly in the presence of air (for instance, aluminum, copper, titanium), exhibit remarkable stability and can only be eliminated through specialized methods like plasma cleaning, in situ sputtering, or chemical etching.

Preparing surfaces before joining is not merely a suggestion; it is frequently an essential requirement for ensuring dependable micro-joining. In industries characterized by high reliability, including aerospace, semiconductor manufacturing, and medical device fabrication, rigorous protocols are adhered to in order to guarantee immaculate surfaces prior to bonding. Potential methods encompass ultrasonic cleaning in solvents such as isopropyl alcohol or acetone, chemical cleaning utilizing dilute acids or bases, and plasma treatment aimed at eliminating organic contaminants and activating the surface. For example, oxygen or argon plasma is frequently utilized to clean polymer and metal surfaces by eliminating carbonaceous layers and boosting surface energy, which in turn improves wetting behavior. In MEMS packaging and wafer bonding applications, surfaces are frequently treated in a cleanroom environment and subsequently joined under vacuum or inert atmosphere to avoid recontamination. Nonetheless, maintaining uniformity across various components, particularly in a high-throughput production environment, presents significant technical challenges. Identical results across various materials may not be achieved under the same process conditions. Minor differences in cleaning time, chemical concentration, or plasma power can result in changes in surface condition, which in turn affects joint quality.

A further technical challenge emerges in ensuring cleanliness during the interval between the cleaning procedure and the subsequent joining operation. After cleaning, surfaces exhibit a high level of reactivity and are susceptible to reoxidation or the adsorption of airborne contaminants, including hydrocarbons or water vapor. In controlled settings, it is crucial to minimize the time interval between cleaning and joining to avoid any degradation of the surface. This presents significant challenges in batch manufacturing settings or when components need to be moved between different pieces of equipment. Furthermore, the surface topography is of significant importance. Rough or porous surfaces have the potential to retain contaminants even post-cleaning, necessitating the use of more intensive techniques like ion beam cleaning or chemical–mechanical polishing (CMP) to achieve adequate cleanliness. Nonetheless, these treatments could potentially modify the surface morphology, introduce additional stresses, or influence the dimensional tolerances of micro components. Consequently, it is essential to find a balance that ensures sufficient cleanliness while maintaining the microstructural and dimensional attributes required for high-precision joining.

Moreover, there is a lack of strong, real-time diagnostics to assess surface cleanliness at the micro and nano scales. Conventional inspection techniques, including visual assessment or contact angle measurement, fall short in identifying contaminants at the monolayer level or thin oxide films. Techniques such as X-ray photoelectron spectroscopy (XPS), Auger electron spectroscopy (AES), or time-of-flight secondary ion mass spectrometry (ToF-SIMS) can provide chemical composition data with high surface sensitivity, but they are expensive, time-consuming, and destructive. Consequently, process control frequently relies on indirect metrics or presumed repeatability, presenting a potential risk for quality assurance in essential applications. Ultimately, this requires the incorporation of process analytics and predictive modeling into surface preparation protocols. Moreover, the cleaning approach should be customized to suit the particular material system being examined. The methods that are effective for gold-silicon bonding could prove harmful to copper-polyimide systems, owing to variations in chemical reactivity, surface energy, and thermal behavior. The unique nature of this specificity poses challenges for standardization and restricts the portability of processes across various platforms or product lines.[30–33]

5.7.5 Joining of Ultra-Thin and Delicate Components

Ultra-thin, fragile components bonding brings a set of special technical difficulties in micro-joining as a result, especially in the areas of flexible electronics, wearable sensors, microfluidic devices, implantable biomedical devices, etc. The devices in question - which can be in the form of thin films, foils, wires, or membranes—have dimensions ranging from tens of micrometers to the nanometer scale. However, due to their small size and low mechanical strength, these substances are likely to be physically damaged when assembling. For example, thin copper foils used in microelectronic interconnects or polyimide based substrates used in flexible circuits are easily damaged, torn, wrinkled, and deformed when subjected to excessive mechanical pressure or thermal stress. Tiny misalignment of the tools or the application of the force to be impaired in any particular location, e.g., during welding, bonding or soldering, can result in the creation of the holes, punctures or stress-cracking of the modified composite structure. The existence of such flaws can have serious consequences: With attempts to rework or repair impossible at these micrometric scales, the whole assembly would have to be replaced.

Thin materials have low-thermal inertia and can be rapidly heated or cooled during joining from a thermal point of view. Such a localized overheating, melting or delamination of, in particular, polymer or polymer-metal laminates may constitute a safety concern. If we deal with micro-joining processes, such as thermosonic bonding or reflow soldering, the heat supply has to be exact to avoid damaging sensitive neighboring structures and to create reliable connections. If they are heated, thin metal wires or foils generate grain coarsening, oxidation or intermetallic compound formation, which may ultimately result in the loss of mechanical strength or electrical conductivity. The mismatch of coefficient of thermal expansion (CTE) of successive layers in the thin multimaterial laminate may also cause wrinkling or buckling of the laminate as it cools, resulting in geometric distortion and debonding at the interface. Such thermal effects need to be carefully controlled by dedicated energy delivery systems, including pulsed lasers, micro heaters or ultrasonic actuation. However, the control system and feedback loops required to achieve such a high degree of accuracy prove to be a rather high technical challenge, especially in high-speed production.

Mechanically speaking, many of the thin parts are too flexible or too brittle to handle, align, and fixture during welding. Silicon thin-wafers are susceptible to breakage under the action of very small mechanical shocks, and thin polymer membranes can deform under vacuum or clamp-induced forces. Micro-manipulators and vacuum-gripping systems are used to limit direct interaction and ensure a uniformly distributed exertion of force. However, these systems have their own limitations, including the potential for contamination, electrostatic discharge (ESD), and difficulties in obtaining accurate placement with high throughput. It is important that the pressure distribution is uniform when connecting non-planar or microstructured components. Specialized conforming tooling or elastomeric interfaces have been designed to address this problem, but these require precise calibration and are often not compatible across different material systems. Furthermore, the consistency of these solutions in production-scale environments remains a focus of ongoing investigation and advancement.

A key issue in the assembly of ultra-thin components is ensuring the durability of joints over time, especially when subjected to mechanical, thermal, or electrical stresses. Thin joints exhibit a higher susceptibility to the initiation and propagation of cracks, primarily because of their restricted ability to absorb energy or redistribute stress effectively. This holds particular significance in environments characterized by dynamic or cyclic loading, like those found in flexible displays, robotic sensors, or implantable cardiac devices. Minor defects or inclusions at the interface—often undetectable during the manufacturing process—can act as sites for crack initiation that may expand under operational conditions. Thus, guaranteeing the reliability of joints in thin structures necessitates not just accuracy during the joining procedure but also thorough post-process inspection and testing. Conventional nondestructive testing (NDT) techniques, including X-ray and ultrasonic testing, frequently fall short in their ability to identify submicron defects in thin interfaces due to resolution limitations. The development and adoption of high-resolution inspection methods is essential, including techniques like acoustic microscopy, nano-CT scanning, and infrared thermography, each presenting unique trade-offs in terms of cost, resolution, and accessibility.

Ultimately, the task of connecting fragile components is closely connected to the advancement of materials, since not every thin material is created with the ease of joining as a priority. Novel materials like 2D structures (graphene, MoS_2), conductive polymers, or bioresorbable metals frequently demonstrate outstanding functional characteristics yet struggle with joinability when employing conventional techniques. These materials might deteriorate when exposed to heat, show a lack of metallic bonding characteristics, or demonstrate poor interfacial adhesion. Consequently, investigations are underway into innovative joining techniques, including laser sintering of nano-inks, low-temperature conductive adhesives, and hybrid bonding strategies that integrate mechanical interlocking with van der Waals or hydrogen bonding. Although these methods show potential, they remain in the research phase and have not yet achieved broad acceptance in the industry. The field currently finds itself at a critical point where innovative approaches that span materials science, mechanical engineering, and microfabrication are essential for effectively connecting ultra-thin and delicate materials within high-performance microsystems.[30,34–37]

5.7.6 Process Monitoring and Quality Assurance in Micro-Joining

The intricacies of process monitoring and quality assurance pose significant technical challenges in micro-joining, primarily due to the small scale of the components and the exceptional precision demanded at every stage of the joining process. Unlike macro-scale welding or assembly, micro-joining does not allow for a consistent margin of error; even the smallest deviations in temperature, pressure, alignment, or time can lead to a failed joint. This poses significant challenges in fields like micro-electromechanical systems (MEMS), advanced microelectronics, medical implants, and optoelectronics, where the performance of the entire device can depend critically on a few sub-millimeter or micrometer-sized connections. The complexity is heightened by the challenges associated with real-time identification and assessment of these discrepancies. Traditional quality control techniques—like tensile testing, visual inspection, or standard electrical resistance measurements—tend to be either destructive, insufficiently precise, or unable to identify submicron voids, cracks, or interfacial defects. The industry is evolving toward advanced, nondestructive, and in-line inspection techniques that ensure dependable process feedback while maintaining the integrity of components.

Monitoring processes in real-time is crucial in micro-joining to guarantee uniform energy distribution and the quality of joints. In various joining techniques such as laser welding, ultrasonic bonding, and thermocompression bonding, the actual duration of the process can range from mere milliseconds to a few seconds. During this limited period, it is essential to meticulously manage and document parameters including energy input, bonding force, displacement, acoustic emission, and temperature gradients. To effectively monitor these signals and detect anomalies throughout the process, it is essential to utilize advanced sensors and data acquisition systems. In the context of ultrasonic wire bonding, it is crucial to monitor bonding parameters in real-time—specifically ultrasonic power, tool velocity, and bond deformation—to ensure that appropriate plastic deformation and intermetallic formation have taken place at the bond interface. Nonetheless, even with progress in sensor technology, interpreting these data streams in real time continues to be a challenging endeavor. Noise, variability in material properties, and tool wear can obscure essential signals or result in false positives and negatives, leading to reliability challenges in feedback-based control systems.

One particularly significant technical barrier that remains is the development of NDE techniques suitable for nondestructive, high-resolution testing of micro-scale joints (COMAP). Traditional approaches such as X-ray imaging or scanning acoustic microscopy (SAM) are used routinely for this purpose but often lack the required resolution for early stage defects like nanometer sized voids, grain boundary separations, or interfacial delaminations. Moreover, such procedures normally precede the initial production, which requires removing parts from the production line, resulting in increased costs and lead time. High-resolution X-ray computed tomography (nano-CT), optical coherence tomography (OCT) and SEM provide detailed information, but the associated costs, low throughput, and lack of possibilities for an in-line monitoring are a drawback. Infrared thermography analysis for thermal mismatch detection and laser ultrasonics for subsurface void identification show promise for rapid, noncontact defect detection. However, these methods are still exploratory, and need to be further developed before they become transferable and scalable to on a high-throughput production-level scale.

Statistical process control and machine learning approaches are becoming increasingly popular for micro-join process data analysis. Its pattern recognition and predictive analytics capabilities help users spot anomalies and trends that indicate a loss of process stability or tool performance. A progressive drift in the bond strength or an increase in the bond variability, for example, can suggest tool wear, contamination or alignment issues which cannot be easily detected by conventional visual inspection means. Artificial intelligence (AI)–based controllers are being researched which modulate the parameters on a real-time basis by tracking the incoming signals continuously. However, such systems are difficult to implement. The requirements of a lot of process-specific training data make the models very time and cost consuming. Furthermore, the opacity of some machine learning models raises traceability, auditability, and regulatory adherence concerns in safety-critical areas, such as aerospace or biomedical device manufacturing.

The lack of universally recognized standards for the quality of micro-joining presents a significant challenge. In contrast to traditional welding and soldering, which adhere to established standards like AWS D1.1 or IPC J-STD-001, micro-joining is characterized by a notable absence of thorough guidelines for assessing joint integrity, defining failure criteria, and conducting qualification testing. This compels every industry or organization to create unique metrics for acceptance, frequently grounded in empirical data and customer requirements. The lack of uniformity obstructs interoperability, quality benchmarking, and adherence to regulations across various sectors. Ongoing collaborative efforts from international standards organizations such as ASTM, ISO, and IEEE aim to address this gap; however, there is a pressing need for broader industry involvement and enhanced data sharing to facilitate robust standard development.

In conclusion, the monitoring of processes and assurance of quality in micro-joining necessitate a fundamental change from conventional inspection and testing methods to cohesive, data-informed, and exceptionally adaptable systems. The intricate nature of the micro-joining environment—defined by extremely small dimensions, rapid processing speeds, and interactions among multiple materials—necessitates the use of sophisticated metrology tools, smart control algorithms, and highly accurate instrumentation. Advancements in this domain will hinge on the effective integration of material science, electronics, software engineering, and AI. Manufacturers can attain the high reliability, yield, and reproducibility necessary for next-generation micro-devices only through this kind of interdisciplinary integration.[4,9,38–40]

5.7.7 Formation of Intermetallic Compounds in Micro-Joining

The creation of IMCs at the joint interface poses a notable technical challenge in micro-joining, particularly in systems that include dissimilar metal combinations like Au-Al, Cu–Sn, or Ni–Sn. IMCs are generally characterized by their hardness, brittleness, and unique chemical properties, which arise from the interdiffusion and chemical interactions between the materials being joined, influenced by thermal or mechanical energy. A thin, controlled intermetallic compound layer can enhance bond formation and mechanical anchoring; however, excessive or uncontrolled growth of these compounds significantly undermines joint reliability, mechanical strength, and electrical performance. This challenge is amplified in micro-joining because of the small dimensions of the components involved. At the micro-scale, a nanometer-thick IMC layer can represent a substantial portion of the overall joint thickness. The growth kinetics and morphology of IMCs require meticulous monitoring and control to guarantee the structural integrity, long-term durability, and functional performance of the joint.

In micro-soldering applications, especially in flip-chip and wafer-level packaging, the development of IMCs like Cu_6Sn_5 and Cu_3Sn frequently occurs at the Cu–Sn interface. While these compounds are essential for creating metallurgical bonding, excessive growth beyond an optimal thickness may result in embrittlement and the initiation of cracks under thermal cycling or mechanical stress. The kinetics of intermetallic compound formation is influenced by various factors such as temperature, time, and diffusion coefficients, as well as the presence of impurities or interfacial defects. In processes such as thermocompression bonding or transient liquid phase (TLP) bonding, higher temperatures are employed to enhance diffusion and reaction rates, subsequently leading to swift IMC growth. This situation is especially harmful in high-density interconnects, as the entire solder joint can evolve into IMCs over time, leading to rigid, brittle areas that exhibit inadequate fatigue resistance. Moreover, the mechanical incompatibility between the IMCs and the base materials can result in residual stresses, which may cause delamination or fracture at the interface. In the Au-Al bonding system, the formation of Au_4Al and $AuAl_2$ IMCs is extensively documented as a contributor to "purple plague," a failure mode marked by the development of voids and the lifting of bonds.

The anisotropic and frequently irregular morphology of IMCs introduces an additional layer of complexity. Under varying process conditions, IMCs can develop into columnar grains, scallop-shaped layers, or needle-like structures, each of which affects crack propagation behavior when subjected to load. For example, scallop-shaped Cu_6Sn_5 grains can serve as stress concentrators, especially at the grain boundaries where voids or Kirkendall porosity might form. Furthermore, the asymmetry in diffusion between the two metals may lead to the formation of voids on one side of the interface, which can further compromise the integrity of the joint. Advanced characterization techniques, including transmission electron microscopy (TEM), SEM, and electron backscatter diffraction (EBSD), have been utilized to investigate the crystallography and development of IMCs at the nanoscale. The findings indicate that even minor variations in thermal history or surface roughness can significantly impact the morphology, orientation, and mechanical properties of the IMC layer. Therefore, managing the thermal budget during the joining process is crucial, as it not only facilitates bonding but also limits the unregulated development of unwanted phases.

Process engineers frequently strive to address IMC-related challenges by optimizing bonding parameters, including temperature, dwell time, and applied pressure. In applications such as ultrasonic welding or wire bonding, utilizing lower ultrasonic power and minimizing bonding time can effectively decrease thermal input, which in turn helps to restrict the formation of IMCs. In a similar vein, employing barrier layers (such as Ni, Ti, Cr) is a prevalent approach to slow down interdiffusion and restrict IMC growth to targeted areas. Nonetheless, the performance of these layers is constrained by factors such as their diffusivity, mechanical integrity, and durability over extended periods of cyclic thermal stress. In emerging processes like solid-state micro-welding, there is an exploration of the use of lower energy inputs and localized heating methods—such as laser micro-welding and micro-resistance welding—to minimize IMC formation altogether. Nonetheless, these processes exhibit a significant sensitivity to variations and necessitate stringent control over input parameters. A further intriguing strategy involves the incorporation of interlayer materials aimed at producing stable yet ductile reaction products, as opposed to brittle IMCs. Utilizing In–Ag systems can result in the development of softer IMCs that exhibit improved compliance and fatigue resistance; however, this advantage may be offset by increased material costs or greater process complexity.

The enduring reliability of micro joints significantly relies on the stability of IMCs in operational environments. In fields like automotive electronics and aerospace sensors, joints frequently experience extended periods of high temperatures and thermal cycling. Under these conditions, IMCs may persist in their growth, a phenomenon referred to as aging, leading to the thickening of the intermetallic layer and an increase in brittleness. The behavior associated with aging is frequently dictated by parabolic growth laws, which suggest diffusion-limited kinetics. In the absence of suitable control measures, IMC layers that were once harmless can transform into major sources of failure following extended usage. Consequently, the application of predictive modeling for IMC growth, utilizing computational thermodynamics (CALPHAD) and finite element simulations, is becoming more prevalent in the design of more resilient micro-joining systems. These models enable professionals to assess IMC thickness over time and across different service conditions, facilitating improved process planning and life cycle management. Even with these advancements, finding the perfect equilibrium between adequate metallurgical bonding and limited intermetallic compound formation continues to pose a significant challenge in both research and industry.[9,30,41–44]

5.8 OPPORTUNITIES IN MICRO-JOINING

5.8.1 Integration in Miniaturized Devices and Electronics

Miniaturization of devices is a growing trend, and such a trend requires precision detecting and micro-joining technology, especially in consumer electronics, telecommunications, and automotive sensor applications. Implantable devices, e.g., invites, sensors and microprocessors are required to have complex assemblies requiring micro-scale connections. For example, traditional methods of bonding, e.g., extensive soldering or mechanical fasteners, generally fail to be viable for such scales due to thermal stress, material incompatibility, or scale constraints. The use of micro-joining methods for attaching structures upon the electrical connections becomes increasingly important for high-density interconnects of micro electronic devices that demand preserving their functionality.

A main advantage of these methods is the shallow HAZ and precise energy deposition, necessary to avoid damaging fragile microelectronic devices. Thermosonic wire bonding is a well-established interconnect technique in the integrated circuits and chip packaging industry. This bonding is achieved by applying heat, ultrasonic vibration, and pressure between the gold or aluminum wire and the die pad and lead frame. Such a concept is highly reliable, and it is widely used for the packaging of ICs (integrated circuits) for smart sensors/controllers.

In addition, the importance of low-temperature bonding technology is increasing due to the increase in heat-sensitive materials, such as polymer and flexible substrate. Methods such as nanosoldering with metal nanoparticle inks and photonic sintering pave the way to produce successful joints at below 200°C, which expands the application scalability of micro-joining in flexible electronics, bio-integrated sensors and transparent conductive films.

The continuous development of system-in-package (SiP) and 3D integration (i.e., stacking multiple chips and components in the vertical direction) requires the improvement of micro-joining technologies. Such technologies must support finer pitches, lower heat dissipation, and joint resistance. In this realm, micro-bump bonding with Cu-pillars/microsolder balls has been identified to have potential in achieving high I/O density with better electrical and thermal performance.[45,46]

5.8.2 Biomedical Device Fabrication and Bio-MEMS

Micro-welding is vital to the manufacturing of implantable medical devices, biosensors, microfluidic chips, and lab-on-chip systems in the biomedical engineering sector. Such devices usually contain an assortment of different materials: soft ones such as polymers, glass, or silicon, or hard ones like titanium or biocompatible ceramics, each of them to be bonded without losing its mechanical and/or biological properties. Unlike traditional manufacturing, biomedical applications need characteristics, including biocompatible, hermetic, and micro-scale precision. These can be reached by advanced micro-joining technologies, such as laser micro-welding, plasma bonding, and diffusion bonding.

For instance, the implantable sensors used in glucose monitoring, cardiac pacing, and neurostimulation require precision adhesion to bond the electronic components with a biocompatible encapsulant shell. Sensor packages, which are typically constructed from materials such as PEEK or titanium, use laser joining to provide hermetic sealing in a manner that is both dimensionally accurate and obviates the use of adhesives that can degrade or create deleterious byproducts. Ultrasonic welds are also used to seal the polymeric microfluidic channels in diagnostic cartridges, to ensure they are free from leaks.

An excellent example of this is the development of bio-MEMS devices which integrate mechanical and electrical functionality on the scale of a micron. Such devices, which include drug delivery and organ-on-chip platforms, rely on bonding methods that preserve micro-channel integrity and cleanliness of the surface. Techniques such as low-temperature plasma-assisted bonding or anodic bonding are preferred for merging glass and silicon layers in microfluidic structures.

The need for minimally invasive medical instruments drives the development of assembly methods that enable smooth surface transitions and strong mechanical connections in microsurgical tools, stents, and micro-needles. Further development of personal and remote health monitoring will depend heavily on the integration of flexible and stretchable electronics with human tissues in the future, and micro-joining of electronic devices to bioresorbable substrates will be very important.[47,48]

5.8.3 Sustainability and Green Manufacturing in Micro-Joining

In the realm of environmentally responsible manufacturing, micro-joining is essential, as it fosters sustainable practices via low-energy processes, enhances material efficiency, and removes toxic substances. Numerous micro-joining techniques, including laser welding, ultrasonic welding, and photonic sintering, provide accurate energy localization. This leads to lower power consumption and minimal thermal losses, especially in contrast to conventional brazing or bulk welding methods. The appeal of micro-joining technologies lies in their potential to help manufacturers minimize their environmental impact.

A significant development is the transition to lead-free soldering and low-temperature joining, influenced by environmental regulations like the Restriction of Hazardous Substances (RoHS) Directive in the European Union. Conventional lead-based solders, while functional, present considerable risks to both the environment and public health. Options such as tin-silver-copper (Sn-Ag-Cu) alloys and nanoparticle-based solders provide enhanced performance while adhering to environmentally friendly standards. Additionally, the use of photonic curing for conductive inks enables rapid, solvent-free integration of printed electronics, significantly minimizing the reliance on harmful chemicals and emissions.

Micro-joining plays a significant role in the conservation of materials. In micro-additive manufacturing, connections are established exclusively in necessary locations, utilizing minimal quantities of joining material. This stands in opposition to bulk joining processes, which frequently encompass extensive interface areas and surplus filler material. The incorporation of recyclable and bioresorbable materials in electronic packaging and wearable devices requires joining techniques that are reversible, biodegradable, or energy-efficient, highlighting the significance of micro-joining in sustainable technologies.

Furthermore, numerous advanced energy devices, including solar cells, microbatteries, and energy-harvesting sensors, depend on micro-joining for the assembly of thin films, interconnects, and microelectrodes. These devices frequently incorporate uncommon or repurposed materials, and meticulous joining is essential to guarantee durability and recyclability. As industries face mounting pressure to embrace circular economy principles, the trajectory of sustainable manufacturing will be intricately linked to advancements in micro-joining.[49–51]

5.8.4 Micro-Joining in Aerospace and Defense Microfabrication

The aerospace and defense industries require unparalleled performance and reliability from each component, especially those functioning under extreme conditions like high G-forces, radiation, vacuum, or cryogenic temperatures. Micro-joining technologies play a crucial role in the production of miniature propulsion systems, microsatellites, infrared sensors, navigation chips, and missile components, all of which demand compact, lightweight, and durable connections.

In the field of aerospace, micro-joining facilitates the lightweight assembly of materials such as titanium, aluminum, Inconel, and composites, which guarantees high strength-to-weight ratios along with enhanced resistance to oxidation and corrosion. Methods like laser beam welding, electron beam welding, and solid-state diffusion bonding are commonly utilized to connect precision components such as micro-gyroscopes, fuel system nozzles, and telemetry equipment.

In defense applications, there is an increasing utilization of micro-scale detonators, MEMS-based fuses, and radar signal processors, all of which necessitate secure, hermetic bonding. Given the potential for catastrophic failures, joining techniques are required to adhere to stringent military standards (MIL-SPEC). Ceramic-to-metal and glass-to-metal seals, created through brazing or laser joining, maintain thermal and mechanical integrity amid varying pressures and temperatures.

Moreover, the emergence of CubeSats and small spacecraft has heightened the need for efficient power and communication systems. This approach utilizes micro-joining techniques for the interconnection of solar cells, antennas, and control circuitry. The application of friction micro-welding, anodic bonding, and micro-riveting has broadened owing to their effectiveness in joining dissimilar materials and facilitating micro-assembly.[52,53]

5.8.5 Flexible and Stretchable Electronics Applications

The emergence of wearable technologies, smart textiles, and soft robotics highlights the critical role of micro-joining in facilitating the advancement of flexible and stretchable electronic systems. In contrast to inflexible substrates, these systems necessitate joining methods that can preserve electrical continuity and mechanical integrity while undergoing bending, twisting, and stretching. Conventional solders or adhesives exhibit brittleness, rendering them inadequate for these specific applications. Consequently, techniques like laser sintering of conductive inks, anisotropic conductive film (ACF) bonding, and ultrasonic spot welding are currently under investigation.

Stretchable electronics generally comprise metallic nanowires, conductive polymers, or carbon-based conductors integrated within elastomers such as PDMS or TPU. It is essential that these materials are connected to microprocessors, sensors, or power sources in a manner that ensures low contact resistance and high durability. Methods such as nano-welding with silver or copper nanoparticles enable the creation of low-temperature, high-strength connections that maintain conductivity even when subjected to deformation.

Innovative micro-joining techniques encompass self-healing junctions, enabling materials to autonomously restore electrical connectivity following mechanical damage. Moreover, the techniques of direct ink writing and aerosol jet printing for conductors facilitate patterned deposition, which can be subsequently enhanced through localized laser or thermal curing for circuit integration on wearable substrates.

Flexible micro-joining finds its utility across various sectors, including healthcare with devices like ECG patches and glucose monitors, consumer electronics featuring foldable phones, and industrial automation through soft sensors in robotic grippers. The long-term viability and commercialization success of these systems will hinge on the reliability and scalability of their joining methods.[54,55]

5.9 EMERGING TECHNOLOGIES IN MICRO-JOINING

5.9.1 Laser Micro-Joining and Ultrafast Laser Techniques

Laser micro-joining as an advanced and universal process has been developed for joining of the micro-scale components with the benefits of noncontact energy delivery, high precision and less thermal damage. This method works by focusing an intense laser beam on a small spot such that materials melt or weld together. It is especially perfect for end lap welding of metals, polymers, ceramics or their composition. Ultrafast lasers (which were first commercialized for material processing in the early 1990s) operating at picosecond or femtosecond time layers, have opened up micro-joining to new frontiers with the capability of cold joining or when the minimum heat diffuses out of the bond interface.

A unique advantage of ultrafast lasers is their ability to bond transparent materials, e.g., glass or sapphire, to metals. This is achieved through nonlinear multiphoton absorption, enabling the structures to be internally modified and micro-welded, while no damage is generated on the surface. These abilities are important for photonics, optical communication and biomedical packaging. As an example, femtosecond laser bonding process is already used in the fabrication of endoscopic devices and high power LEDs, being fully transparent and clean optics is essential.

A significant progression is the application of laser-induced forward transfer (LIFT), and to alongside also laser sintering of nano-inks for flexible electronics, enabling direct-written functionalities and on-the-fly material joining on flexible or curved surfaces. The methods have advantages in terms of being more ecologically friendly, speed, and integrate better to the industry 4.0 type automation concepts.

In addition, recent development of laser hybrid joining, i.e. simultaneous approach of laser and other energy sources like ultrasonic, induction or arc, which has enabled the penetration and wetting of complex material systems like polymers to metals or metal to ceramic joints. Such hybrid structures exhibit a superior strength, fatigue life, and reliability, in particular in aerospace and automotive electronics pheres.[18,56,57]

5.9.2 Nanojoining and Electron Beam-Induced Joining

As the lines between microelectronics and nanotechnology increasingly converge, the rise of nanojoining technologies has become ever more essential. Nanojoining involves the formation of mechanical and electrical connections at scales smaller than one micrometer. This process is crucial for the production of nanosensors, components for quantum computing, nanoelectromechanical systems (NEMS), and molecular electronics. At these scales, conventional joining techniques are not feasible because of limitations related to tool size, excessive heat generation, or the potential for contamination. Consequently, techniques such as electron beam-induced deposition (EBID), FIB welding, and self-assembly-based nanojoining are currently under investigation and application.

The process of joining through electron beams is remarkably efficient for the deposition of conductive nanostructures and the bonding of nano-objects with exceptional nanoscale accuracy. This technique involves directing a concentrated beam of electrons onto a substrate while introducing a precursor gas, leading to localized decomposition and the deposition of the intended material. This method is commonly employed for linking carbon nanotubes to microelectrodes, creating nanowire interconnects, and developing MEMS/NEMS prototypes. These applications require highly precise positioning and pristine processing environments, which EBID delivers.

A further method, diffusion bonding at the nanoscale, facilitates the connection of ultra-thin films (such as gold-to-gold or copper-to-copper) by employing surface activation and pressure at low or moderate temperatures. This holds particular significance in quantum devices and superconducting systems, where the preservation of material purity and joint coherence is essential. The bonding interfaces in these devices should exhibit low resistance and minimal influence on quantum coherence times, rendering nanojoining essential.

Furthermore, the use of metal nanoparticles as intermediate layers in joining processes, which sinter at lower temperatures, is becoming increasingly favored in the realm of flexible and stretchable electronics. The observed melting point depression in these nanoparticles, attributed to size effects, facilitates the low-temperature joining of heat-sensitive substrates. Obstacles like beam-induced damage, elevated equipment costs, and sluggish deposition rates continue to impede broad implementation; however, progress in automation and beam control technologies is expected to address these challenges soon.[58–60]

5.9.3 AI-Based Process Control and Digital Twin Integration in Micro-Joining

The combination of AI and digital twin framework in micro-joining bestows outstanding precision, automation, and prediction potential. Micro-joining is strongly influenced by different parameters, such as material combination, joint shape, and surrounding environment. Variation of these parameters leads to defects like voids, cracks, or poor bonding. AI-based systems feed off live data inputs—whether these documents are thermal, acoustic or optical sensors—in order to precisely calculate, monitor, and control the science of "joining."

AI-powered systems using supervised machine learning algorithms can analyze historical joining data and forecast process results such as joint strength, defect formation, and bonding uniformity. In ultrasonic welding, methods based on modeling of the vibration signature and energy profile have predictive models that can predict weld quality and adapt the parameters on line to prevent failure. This significantly reduces the amount of manual adjustments and checks, resulting in increased production efficiency and uniformity.

The concept of digital twins—a virtual replication of the live joining system—is fast becoming established in high-tech manufacturing. The twins simulate joining operations in a virtual environment using real-time sensorics, CAD models, and simulations of materials behavior. Using digital twins in micro-joining, thermal gradients and stresses and the performance of a joint can be predicted, and engineers can improve process parameters before performing real trials. This reduces experimental costs and shortens the time from design to manufacturing.

Furthermore, the combination of digital twin technologies with AI enables adaptive process control, in which the joining equipment can autonomously react to a disturbance, the changing material properties, or wear of the tool. This is especially beneficial in environments featuring many different kinds of products in smaller volumes, which usually require individualizing settings for each batch. In the era of Industry 4.0 and the era of intelligent factories, the incorporation of AI-based micro-joining systems will be of great importance in the cyber-physical production systems (CPPS) to achieve more agile, traceable, and fault-tolerant manufacturing activities.[61–63]

5.10 FUTURE TRENDS IN MICRO-JOINING

5.10.1 Integration of Artificial Intelligence and Machine Learning in Micro-Joining

The integration of AI and machine learning (ML) is profoundly transforming the future of micro-joining, particularly in areas such as process design, monitoring, and control. These technologies facilitate immediate forecasting, flexible management, and enhancement of joining parameters through insights gained from extensive datasets produced during joining processes. In the realm of laser micro-welding, algorithms are being developed to establish connections between process parameters (including pulse duration, energy, and frequency) and metrics of weld quality (such as weld strength, porosity, and dimensional accuracy). This predictive ability minimizes dependence on costly trial-and-error experiments, thereby improving productivity and process efficiency. Deep learning models, particularly convolutional neural networks (CNNs), are increasingly utilized in vision-based inspection systems for the automatic detection and classification of defects at micro-scales.

Furthermore, digital twins powered by AI are being developed for joining processes, enabling virtual prototyping and simulations that replicate the system's real-time responses under different conditions. This enables flawless manufacturing by utilizing predictive diagnostics and prescriptive maintenance strategies. These digital frameworks hold significant potential for intricate microelectronic assemblies, where maintaining high standards of quality control is essential. Real-time decision-making systems supported by reinforcement learning have the capability to dynamically adjust joining parameters during operation, allowing for adaptation to changes in material properties or surface conditions. With advancements in data acquisition and sensor technology, AI/ML systems are set to significantly enhance the intelligence, reliability, and autonomy of micro-joining processes.[64,65]

5.10.2 Adoption of Ultrashort Pulse Laser and Hybrid Laser Joining Techniques

The future for micro-joining looks promising, especially with the increased utilization of ultrashort pulse lasers, such as femtosecond and picosecond lasers, in hybrid laser joining processes. Ultrashort pulse lasers allow for super selective energy deposition with minimal HAZ help to restrict thermal damage to a minimum, and uphold the capability for micro- and nanoscale assembly. This is of great relevance for dissimilar materials, as for instance metals with polymers or ceramics, which are difficult to weld because of their different thermodynamic properties. In addition, the nonlinear absorption process in the ultrashort pulse laser induces selective material interaction and increases control of the melt pool, and, therefore, the mechanical and electrical properties of the bond.

Moreover, the hybrid processes such as laser-ultrasonic bonding and laser-assisted micro-friction welding have been developed to make use of the complementary effects of different energy sources. These practices improve tensile and other forms of stress, bend load, and other forms of strength, as well as the reliability and process flexibility of the joint, while promoting the attainment of a joint through lower-temperature joining and reducing residual stresses. Especially in electronics and biomedical devices, which demand a high degree of sensitivity to materials and miniaturization, the hybrid processing offers an opportunity to tailor the heat profiles and the supply of energy, which helped in achieving excellent micro-scale bonding. The next generation of innovation will likely involve equipping ultrashort pulse lasers with adaptive beam shaping, real-time feedback, and AI augmented control, allowing for the vision of what micro-welding can achieve to be pushed further.[66,67]

5.10.3 Development of Environmentally Friendly and Low-Energy Joining Techniques

The focus on sustainability is also taking form in the manufacturing industry with new trends in micro-joining, where cutting down on environmental footprints and energy consumption is a priority. Traditional thermal joining processes usually require high temperature, knives gases and vacuum, resulting in huge energy consumption, environmental pollution, and high operating costs. Low-temperature solid-state, micro-joining methods, such as ultrasonic welding, cold spray joining, and TLP bonding, are progressively being considered as substitutes. The approaches are aimed at saving the energy ultimately and quantitatively cutting down the thermal distortion and residual stress in the fragile micro part.

In addition, research is being conducted on bio-inspired and solvent-assisted micro-joining applications such as biodegradable electronics and biomedical implants. Many of these solutions involve renewable or recyclable materials and are designed for disassembly and the recovery of the material. The drive toward sustainable electronics is fostering innovation in waste-reducing technologies, such as the development of water-soluble solder materials and laser-releasable adhesives. In light of the increasingly stringent environmental legislations and the life cycle analysis that is becoming consolidated, the industry is likely to more and more embrace those eco-efficient joining processes that can be adjusted to deliver minimum carbon footprints and that support the concept of the circular economy. Life cycle analysis (LCA)–supported assessment of joining processes is foreseen as being key for deciding which processes will be chosen and how they will be designed in the future's manufacturing systems.[68,69]

5.10.4 Micro-Joining in Additive Manufacturing and 3D Printed Assemblies

The combination of micro-joining techniques with additive manufacturing (AM) will revolutionize fabrication processes for future multimaterial and functional microdevices. In the case of micro-additive manufacturing (μAM), since AM has moved into the field of μAM, there is a need for joining processes with high accurate, small heat input, and no contamination. Although conventional AM processes can produce complex and feature-rich geometries, it may be necessary to add dissimilar parts or off-the-shelf components (e.g., microelectronic chips, sensors, or actuators) during post-processing. A micro-joining in that heterogeneous integration to convert static printed components into smart functional devices is also necessary.

Techniques such as laser micro-welding, ultrasonic bonding, and thermocompression bonding are being developed for the integration of printed microstructures with embedded electronics and flexible substrates. Interestingly, advances in micro-joining can be used to enable the seamless joining of metals to polymers, ceramics, and biological materials. When hybrid layered materials are 3D printed, micro-joining is an essential factor that determines the mechanical integrity and electrical conductivity of interfaces. Furthermore, the coupling thereof with AI-guided process control and in situ monitoring equipment lead to direct integration of micro-welding within the AM toolpath, thereby making the path for hybrid assembly continuous and automated.

As multimaterial 3D printing, with applications in aerospace, electronics, and medical diagnostics, progresses, the need for micro-joining can be anticipated to be an important part of enabling embedded systems, multilayer circuits, and fluid flow channels, all within a single build cycle. This capacity allows full monolithic assembly and printed microfactories where parts are intricately linked via carefully controlled micro joins.[70,71]

5.10.5 Nanomaterial and Composite Integration in Micro-Joining

The integration of nanomaterials and advanced composites represents a pivotal advancement in the evolution of micro-joining. The growing application of carbon nanotubes (CNTs), graphene, MXenes, metal matrix composites, and FRP systems in miniaturized electronics and aerospace components introduces novel joining challenges. The materials demonstrate distinctive characteristics, including anisotropy, elevated thermal conductivity, and reduced interfacial bonding energy, which render traditional joining techniques ineffective or potentially harmful. Consequently, advanced methods like nano-brazing, laser-induced localized sintering, and field-assisted micro-joining (utilizing electric/magnetic fields) are being innovated to establish stable, high-performance joints at the nano/micro-scale.

The bonding mechanism at nano-interfaces exhibits distinct characteristics compared to traditional metallurgy. The process is characterized by van der Waals forces, chemical functionalization, and mechanical interlocking, as opposed to bulk diffusion. Consequently, the integration of surface activation methods such as plasma treatment, ion beam bombardment, or laser cleaning into the joining process is on the rise. Furthermore, there is a growing development of nanoparticle-based solders and adhesives that cure at reduced temperatures while providing adjustable electrical and thermal characteristics. These materials enhance the ability to rework and allow for joining in environments sensitive to temperature, such as in biomedical implants and wearable electronics.

Upcoming investigations are concentrating on atomic-scale modeling and simulation of the joining interface to forecast joint behavior and enhance material combinations. High-resolution microscopy and spectroscopy for in situ characterization are facilitating real-time analysis of interfacial evolution, phase formation, and nanostructure growth during joining. The insights presented here are essential for the design of joints that possess specific functionalities, including high conductivity, corrosion resistance, and mechanical robustness, aimed at advancing next-generation nanoelectronic and photonic systems.[72,73]

5.10.6 Micro-Joining for Flexible, Stretchable, and Wearable Electronics

Flexible and wearable electronics signify a groundbreaking area in medical diagnostics, consumer electronics, and human-machine interfaces. These systems require micro-joining techniques capable of handling deformation, stretchability, and repeated mechanical cycling without experiencing failure. Traditional soldering or welding methods face challenges such as brittleness, the need for high-temperature processing, and the presence of rigid interconnects. As a result, upcoming trends are moving toward bonding methods that operate at low temperatures and pressures, as well as the use of stretchable conductive adhesives. This process involves the integration of various materials such as elastomers (e.g., PDMS), metallic films, textiles, and polymer composites, ensuring optimal performance even under dynamic strain conditions.

Innovative methods like laser sintering of nanoparticle inks, ACFs, and liquid metal joining are being utilized to develop robust, flexible interconnects. For example, gallium-based liquid metals such as EGaIn are utilized to create self-healing, stretchable connections that exhibit remarkable electrical conductivity. Additionally, conductive traces created through inkjet printing are bonded via photonic sintering or microwave heating, providing swift, targeted energy application while preserving the integrity of the substrates. The compatibility of these processes with roll-to-roll manufacturing enhances their scalability, facilitating the mass production of wearable devices.

Future advancements in micro-joining for wearables will incorporate sensors for strain, temperature, and bio-signals directly within the joint interfaces. This paves the way for advanced joints that not only ensure mechanical connectivity but also serve as functional nodes in data acquisition and communication networks. The rising need for biocompatibility and skin-conformal electronics has led to significant advancements in biocompatible adhesives and joining materials that replicate the softness and resilience of biological tissues.[74, 75]

5.10.7 Micro-Joining for Quantum Devices and MEMS/NEMS Systems

As we move into the quantum computing era and the expansion of micro- and nanoelectromechanical systems (MEMS/NEMS) accelerates, the requirements for micro-joining technologies are experiencing a significant shift. These applications require joining techniques that function at very small scales—often at the atomic or molecular level—while ensuring ultra-high precision, minimal thermal input, and compatibility with sensitive materials such as superconductors, piezoelectrics, and 2D semiconductors. For instance, quantum devices constructed using Josephson junctions or qubit arrays exhibit significant sensitivity to thermal and vibrational disturbances. Consequently, the joining processes for these systems must be designed to minimize interfacial contamination, residual stress, and nonuniformities.

Innovative approaches encompass cryogenic-compatible bonding techniques like low-temperature wafer bonding, eutectic nano-brazing, and van der Waals-based adhesion, which facilitate defect-free interfaces. These elements are essential for ensuring the coherence and operational integrity of quantum circuits. In the realm of MEMS/NEMS fabrication, methods such as FIB welding, localized laser joining, and thermocompression bonding are undergoing refinement for the assembly of micron-scale components, including actuators, micro-cantilevers, and resonators. Atomic layer deposition (ALD)–based joining, characterized by controlled monolayer growth that facilitates seamless interfaces, is increasingly recognized for its capacity to generate hermetic seals and mechanically robust nano-joins with atomic precision.

Anticipating future developments, the incorporation of joining techniques into the microfabrication and lithography workflow will be crucial. The focus is on creating hybrid CMOS-quantum architectures, which involves the precise micro-joining of superconducting materials to traditional semiconductors, ensuring that their functionality remains intact. Cutting-edge process monitoring tools like TEM and synchrotron X-ray imaging are utilized to investigate these ultra-small joints in real time, allowing for the refinement of joining protocols as needed. In conclusion, the micro-joining landscape for quantum and NEMS technologies will be shaped by the principles of quantum-safe processing, atomic-level interfacial design, and extreme environmental compatibility.[76,77]

5.10.8 Self-Healing and Reconfigurable Micro-Joining Interfaces

In the near future, micro-joining is set to evolve from static, irreversible bonds to dynamic, adaptive interfaces capable of self-healing, reconfiguring, and restoring functionality in real-time. Drawing from biological systems, the development of self-healing micro-joining technologies is underway, utilizing dynamic covalent bonds, reversible polymers, and microencapsulated healing agents. These systems hold significant importance in aerospace, wearable, and implantable electronics, where the need for long-term durability and maintenance-free operation is critical. In instances where microcracks or delamination manifest at the joint interface, embedded agents are either released or thermally activated to reinstate the structural or conductive properties of the joint.

Dynamic bonding mechanisms, including Diels-Alder chemistry, hydrogen bonding, and shape-memory polymer interfaces, facilitate the repeated formation and separation of joints without any degradation. This enhances the ability to rework, modularize, and recycle microsystems. In the realm of reconfigurable electronics, the implementation of reattachable joints facilitates the removal, upgrading, or repositioning of modules or chips without incurring mechanical damage. Recent advancements in reversible micro-joining feature the application of anisotropic adhesives that bond selectively in one direction, alongside dynamic metallic nanoparticles that sinter in response to light or magnetic fields and can subsequently dissolve or delaminate as needed.

The adoption of reconfigurable micro-joining is increasingly recognized in the field of soft robotics, where joints need to adapt dynamically to movement, deformation, and variations in operational state. For example, joints that incorporate magneto-responsive materials can be activated from a distance to alter their shape or separate in response to an external stimulus. The integration of self-healing and reconfigurable joining technologies will facilitate the development of intelligent materials that are structurally adaptive and possess the ability for autonomous repair. This advancement will greatly prolong product lifespans and support closed-loop manufacturing practices.[78,79]

5.10.9 Integration of In Situ Sensing and Real-Time Feedback Control

Future micro-joining systems are advancing into sophisticated manufacturing platforms that can monitor and adjust their operations in real time. In situ sensing technologies—such as infrared thermography, acoustic emission sensors, OCT, and Raman spectroscopy—are being integrated into micro-joining workstations to track temperature gradients, surface topography, chemical interactions, and joint formation dynamics. These sensors facilitate the identification of defects, contamination, or thermal anomalies during the joining process, offering a closed-loop feedback mechanism for ensuring quality.

With the integration of AI-driven controllers, these sensing systems possess the capability to dynamically adjust parameters such as laser power, pulse duration, or tool pressure, thereby maintaining consistent quality despite fluctuating conditions. In the context of micro-resistance welding for battery interconnects or bioelectronic circuits, it is possible to utilize sensors to monitor nugget formation and adjust current pulses as needed. This degree of oversight is crucial in scenarios where even a micron-scale flaw could lead to disastrous system breakdowns. Additionally, wireless micro-sensors integrated into components can transmit real-time information regarding joint health, temperature variations, or strain throughout the product's lifespan, facilitating predictive maintenance approaches.

Innovative strategies involve creating "smart tools" that incorporate microcontrollers, edge AI processors, and multimodal sensors, facilitating decentralized control and enhancing process reliability. With the rise in process complexity due to the integration of diverse materials and geometries, the necessity for real-time monitoring and adaptive systems will become crucial. As these technologies evolve, they will facilitate the development of self-correcting micro-joining processes and autonomous assembly systems that enhance their performance during production, thereby increasing yield, reliability, and product lifespan.[80,81]

5.10.10 Micro-Joining under Extreme Environments (Vacuum, Cryogenic, High Radiation)

Micro-joining technologies are required to adapt to increasingly extreme service conditions, including high vacuum environments (such as those found in space systems), cryogenic temperatures (as seen in superconducting systems), and high-radiation settings (typical of nuclear and deep-space electronics). These applications present significant challenges related to thermal expansion mismatch, oxidation, interfacial diffusion, and the degradation of joining materials. Traditional soldering and welding methods frequently prove impractical, highlighting the need for the advancement of strong solid-state techniques like diffusion bonding, glass frit bonding, and thermocompression welding, which utilize specifically engineered interlayers.

In vacuum environments, it is essential for micro joins to be hermetically sealed and to endure outgassing and electron bombardment. This necessitates extremely clean surfaces and vacuum-compatible materials such as gold, indium, and silicon oxynitride for interfaces. In cryogenic applications, it is essential for joints to maintain both mechanical and electrical integrity, even at temperatures as low as 4 K. Investigations are underway into materials that exhibit matched CTE, including beryllium-copper and molybdenum-based alloys, for their potential use in cryo-compatible micro joins. Furthermore, employing amorphous or nanocrystalline joining materials minimizes phase transitions that may lead to joint failure under thermal cycling conditions.

Exploration is underway into joining techniques that are resistant to radiation, utilizing materials such as tantalum-, tungsten-, and ceramic-based adhesives that can withstand such conditions. Ongoing investigations focus on the integration of nanostructured metal foams or metamaterials capable of absorbing radiation and minimizing transmission at the joined interface. The advancement of future space-based microelectronics, satellites, and quantum systems will rely significantly on these innovative joining solutions, which are designed to endure prolonged exposure to space radiation, micrometeorites, and extreme temperature variations.[82,83]

5.11 ARTIFICIAL INTELLIGENCE IN MICRO-JOINING PROCESSES

5.11.1 Overview of AI-Driven Micro-Joining

The integration of AI into micro-joining processes represents a paradigm shift from traditional manufacturing approaches to smart, adaptive, and self-optimizing systems.[84] As micro-joining demands unprecedented levels of precision, repeatability, and quality assurance, AI technologies have emerged as critical enablers for meeting these challenges. The convergence of AI with micro-joining addresses fundamental limitations in human-operated processes, including real-time parameter optimization, defect prediction, and quality monitoring at scales beyond human perceptual capabilities. The adoption of AI in micro-joining is driven by several factors.

5.11.1.1 Complexity of Process Parameters

Power, speed, pressure, temperature, and many other interdependent parameters are involved in micro-joining, necessitating complex optimization. These factors show nonlinear interactions that are difficult for conventional control systems to handle. These intricate relationships may be modeled by AI algorithms, which can also find ideal parameter combinations that are impossible to find by human exploration.

5.11.1.2 Real-Time Quality Requirements

Millisecond-level response times are necessary to prevent faulty joints in higher end components by instantaneous process rectification. AI systems are able to instantly modify joint quality by analyzing process on real-time basis. This skill is especially important in sectors where post-process inspection is impossible, such as the manufacturing of medical devices.

5.11.1.3 Data-Rich Manufacturing Environments

Large volumes of process data are produced by contemporary sensors, which can be used to make wise decisions. Multidimensional data streams containing important process information are produced by high-speed cameras, thermal imagers, acoustic sensors, and plasma monitors. AI systems are able to extract valuable patterns from this flood of data that would be too much for human analysts and traditional analysis techniques to handle.

The need for zero-defect manufacturing is especially important in electronics, aircraft, and medical applications where joint failures can have disastrous outcomes. By spotting minute process variations that come before flaws, AI makes predictive quality control possible. This proactive strategy drastically lowers scrap rates and reworks expenses by shifting quality management from detection-based to prevention-focused.

5.11.2 AI Technologies in Micro-Joining

5.11.2.1 Supervised Learning for Process Optimization

Quantitative correlations between process settings and mechanical qualities can be established using regression models for joint strength prediction based on input parameters. These models drastically reduce development time and material waste by enabling virtual parameter optimization without the need for actual experiments. Robust process design can be supported by advanced ML approaches that offer forecast confidence intervals.[85]

Using process monitoring data, classification algorithms for defect identification and categorization can automatically discover different sorts of flaws. Porosity, cracks, and incomplete penetration can be accurately distinguished using support vector machines and random forests. When serious flaws are found, these systems can initiate automatic process adjustments or equipment shutdowns to stop the bulk manufacturing of defective parts.

Complex interactions between laser power, pulse duration, and material properties can be captured by neural networks for nonlinear parameter relationships in laser micro-welding.[86]

5.11.2.2 Unsupervised Learning for Pattern Recognition

Similar welding conditions and their corresponding results can be automatically grouped using clustering algorithms for detecting similar process regimes. In high-dimensional parameter spaces, natural groupings that human operators would overlook can be found by K-means and DBSCAN algorithms. This feature makes it possible to automatically identify and optimize the process for novel materials with joint configurations.

Anomaly detection employs statistical techniques to discover outlier conditions that could result in problems. One-Class SVMs and Isolation Forests are capable of identifying minute departures from typical functioning that may point to upcoming breakdowns. Early identification of these abnormalities enables preventive measures to be taken before faulty joints are manufactured, thus lowering scrap rates.

5.11.2.3 Reinforcement Learning for Adaptive Control

Self-optimizing systems that get better over time are produced by real-time parameter change based on process input. Policy and Q-learning Control systems that optimize joint quality and minimize energy usage can be developed using gradient approaches. Without human assistance, these systems are able to adjust to shifting climatic conditions and equipment deterioration.[87]

Self-learning systems gain knowledge from each joining operation carried out and get better with operational experience. Strong Reinforcement Novel control strategies that surpass human-developed methods can be found via learning algorithms. These systems can handle new materials and collaborative designs with little retraining or programming work because of continuous learning.

Reward functions that take into account several performance measures in real time are used in multiobjective optimization to balance quality, speed, and energy usage. Control strategies that offer the optimum trade-offs between conflicting objectives can be found using Pareto-optimization techniques. In manufacturing settings where joint quality is just as critical as energy efficiency, reinforcement learning is crucial.

5.11.3 Computer Vision Systems

5.11.3.1 Real-Time Monitoring and Analysis

Joint formation dynamics can be recorded at microsecond timescales using high-speed cameras and image processing methods. During laser welding operations, these systems may monitor keyhole stability, melt pool behavior, and plasma plume properties. Process instabilities that may result in defects can be quickly identified because of the real-time analysis.[88]

Heat distribution during joining activities can be measured as noncontact using thermal imaging for temperature distribution studies. Temperature gradients and cooling rates that affect microstructure development can be observed by infrared cameras equipped with the proper filters. AI systems can predict quality from thermal signatures by correlating thermal history with the final joint attributes.

Spectrometer data is used in spectral analysis for plasma plume characterization in laser welding to track penetration and process stability. Certain spectral lines can indicate excessive evaporation or defect formation, as well as the presence of vaporized alloying materials. An additional monitoring tool is the ability of ML algorithms to learn the correlation between joint quality and plasma spectra.

5.11.3.2 Identification and Categorization of Defects

Joint defects can be automatically classified from macro and micrographs using CNNs for visual defect diagnosis. These networks are more accurate than human inspectors at identifying defects such as undercut, porosity, and cracks. Even with small training datasets tailored to micro-joining applications, efficient performance is made possible by transfer learning from pretrained networks.[89]

Transfer learning uses pretrained models for small datasets to apply knowledge from large image databases to micro-joining applications. Models trained on natural pictures can be improved for weld fault detection with very few labeled instances. This approach significantly reduces the amount of data collection and labeling labor required to develop effective vision-based inspection systems.

Multimodal vision systems that combine visible, infrared, and X-ray imaging provide comprehensive joint evaluation from several perspectives. By merging data from various imaging modalities, data fusion techniques can improve the accuracy of defect identification. Surface indicators that are not visible to surface investigation methods can be correlated with subsurface faults using these technologies.

5.11.4 Digital Twin Technology

5.11.4.1 Virtual Process Replication

Accurate digital representations of micro-joining processes are produced by combining real-time sensor data with physics-based models. Thermal profiles, stress distributions, and microstructure change during joining can all be simulated using finite element models. Accurate joint property prediction is made possible by the constant updating of these simulations using sensor data to keep them in sync with the physical process.[90]

Model-based reasoning is used in predictive maintenance using digital replica analysis to forecast equipment faults. Digital twins can mimic the deterioration of vital parts such as ultrasonic horns, welding electrodes, and laser optics. These systems minimize unplanned downtime by predicting maintenance requirements prior to failures by comparing real performance with simulated ideal behavior.

Risk-free process setting exploration is made possible by virtual parameter optimization prior to physical implementation. In the digital environment, engineers can test extreme parameter combinations without causing harm to materials or equipment. This feature makes it possible to optimize current processes beyond standard operating limitations and speeds up the development of new materials.

5.11.4.2 Hybrid Modeling Approaches

Models that integrate data-driven insights with physical knowledge are produced by combining ML and FEA. While learning from experimental data, physics-informed neural networks are able to enforce boundary restrictions and conservation principles. Compared to strictly data-driven methods, these hybrid models usually require less training data and offer more accurate extrapolation.

The reason why digital twins continue to be realistic approximations of their physical counterparts is thanks to real-time model updating based on process feedback. As new sensor data becomes available, Bayesian updating techniques can modify model parameters to account for material variability and equipment drift. Throughout the equipment lifecycle and across material batches, this adaptive capacity preserves model correctness.

Confidence intervals for model predictions are produced by quantifying uncertainty for reliability evaluation. The effect of input variability on joint quality forecasts can be measured using Monte Carlo techniques and Bayesian inference. This data helps engineers comprehend the dependability of AI-based recommendations in the face of noise and uncertainty and supports robust process design.[91]

5.11.5 Implementation Framework for AI in Micro-Joining

5.11.5.1 Data Infrastructure and Sensor Integration Strategy

For thorough process monitoring, multi-sensor fusion integrates information data from electrical measurements, thermal cameras, sound sensors, and vision systems. These different data streams can be combined into cohesive process representations using Kalman filters and Bayesian networks. When individual sensors malfunction or produce noisy data, this multimodal approach offers redundancy and enhances monitoring dependability.

Process dynamics are captured by high-frequency data collecting devices with enough temporal resolution to observe important occurrences. To manage the massive data quantities produced by the devices, sufficient bandwidth and storage capacity are necessary.

To reduce latency for control decisions, edge computing for real-time processing analyzes data in close proximity to the operation. Neural network inference can be carried out with millisecond latency using specialized AI circuits and field-programmable gate arrays (FPGAs). By processing data locally, this architecture reduces network bandwidth needs and allows for instantaneous process modifications (Figure 5.1).

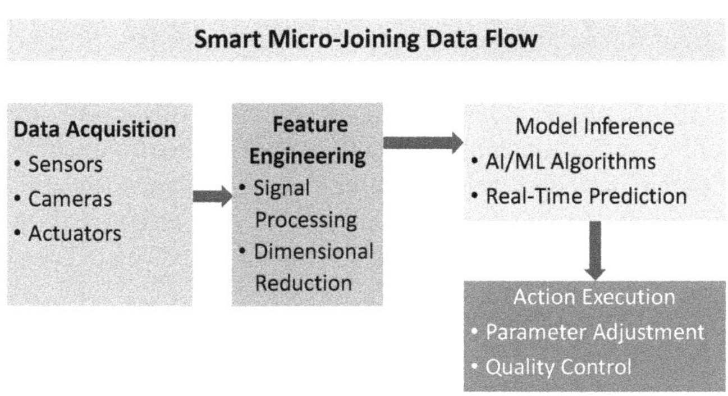

FIGURE 5.1 Data management architecture.

5.11.5.2 Model Development Lifecycle

During training, AI models are exposed to a variety of operational circumstances employing design of experiments (DoE). Ideal experimental designs minimize the number of necessary trials while optimizing information gain. These methodical techniques yield models with greater capacity for generalization and are more effective than one-factor-at-a-time experimentation.

Active learning for effective data gathering carefully selects the most instructive samples for model training and labeling. The system prioritizes parameter areas that have substantial model uncertainty for experimental study. This method maximizes the improvement in model performance while minimizing the experimental cost of data collecting.[91]

By utilizing prior knowledge, transfer learning from related procedures or resources speeds up the creation of models for new uses. With little more data, models trained on one material system can be adjusted for comparable materials. In situations where comprehensive experimental data is not yet available, this feature is very helpful for the development and processing of novel materials.

5.11.5.3 Model Validation and Testing

Model performance is evaluated on unobserved data from the real manufacturing environment by cross-validation using production data. K-fold cross-validation assists in identifying overfitting and offers reliable performance estimates. Instead of relying solely on historical data, proper validation guarantees that models will function dependably when implemented in production.

In controlled production trials, A/B testing compares the performance of new AI models with current control tactics. Key performance metrics can be statistically compared by split testing with random assignment. This method minimizes production disturbance while enabling data-driven model deployment decisions.

New production data is incorporated into ongoing model updates and improvements to sustain model correctness over time. Online learning algorithms are able to adjust to new material batches, equipment modifications, and process drift. As industrial conditions change, regular model retraining promises that AI systems continue to be effective.

5.11.6 Technical and Implementation Barriers

(a) **Technical Barriers**

- Data Availability and Quality – Restricted labeled datasets for uncommon defect categories
- Problems with sensor noise and calibration and synchronization of data from several sources
- Model Generalization – adaptability to process drift and equipment aging and transferability across various materials and geometries
- Managing unfamiliar circumstances outside of training data

(b) **Implementation Barriers**

- Infrastructure and Cost – Integration with existing production systems
- High initial investment in sensors and computer infrastructure
- Skills gap in AI and manufacturing convergence
- Issues with Regulation and Compliance – Requirements for explainability and openness
- Validation of AI systems for regulated industries
- Data security and intellectual property issues

5.12 CONCLUSION

5.12.1 Summary of Key Challenges in Micro-Joining of Materials

Micro-joining is an enabling technology for building the next generation of miniaturized devices in applications ranging from microelectronics, biomedical devices and aerospace components to quantum systems. However, its performance is confronted by a variety of technical challenges that require in-depth scientific and engineering solutions.

One major difficulty is to find materials that are compatible and this is particularly important for joining of different materials, e.g., metals to ceramics or polymers or the like. The CTE, bond energy and chemical reactivity of these materials are largely different. The interactions often result in either thermal stress-induced fractures, the formation of IMCs that are brittle, or interface delamination. A typical example is the Cu-Al or Au-Al systems in which intermetalllics such as Cu_6Sn_5 or Au_5Al_2 are employed, resulting in serious deterioration in both mechanical and electrical reliability.

The details of thermal management in the joining process are remarkable. Due to the small thermal mass associated with a micro-component, temperature changes even on a very small scale may cause undesirable outgassing, microstructural changes, or thermal degradation. This need for precise control over heat input is even more important in hybrid systems, in which organic substrates are coupled to metal interconnects.

The precision and alignment requirements at the micro- and nanoscale are equally critical. A slight deviation of merely a few microns can result in joint misalignment, functional loss in MEMS/NEMS, or signal attenuation in optoelectronic systems. The manipulation of ultra-thin foils or membranes necessitates sophisticated fixturing, precise nano-positioning systems, and actuation that is controlled by feedback mechanisms to prevent tears or deformation.

Surface cleanliness remains a continual challenge. Successful bonding in micro-joining relies on atomically clean surfaces, especially during diffusion or thermo-compression processes. Even the slightest presence of oxide or organic contaminants, down to the monolayer level, can significantly impede wetting or bonding, leading to diminished joint strength or potential electrical failure. Even with progress in plasma cleaning and ion beam surface preparation, there is still a notable absence of real-time cleanliness diagnostics.

The development of IMCs at the joint interface represents a significant challenge to joint reliability, particularly in systems involving dissimilar metals. Although thin intermetallic compound layers can enhance metallurgical bonding, their uncontrolled growth results in brittleness and premature fatigue failure. Given the limited volume of micro joints, even nanometer-scale IMC layers can represent a substantial fraction, significantly impacting joint performance when subjected to mechanical or thermal stress.

Monitoring and quality assurance are still areas that require further development in micro-joining. Conventional nondestructive testing (NDT) techniques fall short when it comes to detecting micro-scale defects such as voids, cracks, or incomplete fusion. Methods such as acoustic microscopy or SEM provide enhanced insights; however, they tend to be costly, time-intensive, and lack in-line capabilities. Consequently, numerous micro-joining processes function with limited feedback, resulting in variability and diminished reproducibility in high-throughput settings.

The integration of ultra-thin and delicate materials presents further mechanical challenges. Thin films or flexible substrates such as PDMS, graphene, or polyimide exhibit a high sensitivity to mechanical stress or localized heating. The application of bonding force should be consistent and precisely adjusted, with specialized compliant tooling frequently required to avoid mechanical failure.

Ultimately, the absence of universal standards for micro-joining practices, material qualifications, and quality control metrics poses a significant obstacle to achieving industrial scalability. In contrast to macro-welding, which adheres to established standards like AWS or ISO specifications, micro-joining techniques differ significantly across various sectors and are frequently reliant on empirical adjustments. This hinders interoperability and wider acceptance in essential sectors.

5.12.2 Highlighted Opportunities in Micro-Joining

In light of these significant challenges, micro-joining offers a diverse array of opportunities that correspond with emerging technological demands, particularly as sectors progress toward ultraminiaturization, sustainability, and multifunctional integration. A particularly promising area is its essential contribution to the miniaturization of electronic and photonic devices. The growing convergence of sensors, controllers, antennas, and power sources into a compact micro platform necessitates the use of highly localized, low-thermal, and precise joining techniques.

Biomedical applications represent a crucial area for expansion and development. The increasing need for implantable devices, lab-on-chip systems, and minimally invasive sensors has led to the integration of various biocompatible materials, including titanium with polymers or glass. Micro-joining techniques like laser micro-welding and plasma bonding have demonstrated effectiveness in creating hermetic, biocompatible, and dimensionally accurate joints for pacemakers, stents, and microfluidic cartridges.

Modern industrial strategies are increasingly centered around sustainability and green manufacturing. Micro-joining technologies provide energy-efficient options compared to traditional joining techniques. The use of lead-free soldering, nanoparticle-based inks, and recyclable conductive adhesives significantly reduces environmental impact. The capability to connect with minimal filler material and nearly zero emissions is essential for the development of sustainable electronics and flexible systems.

In the fields of aerospace and defense, the demand for lightweight, high-strength, and multifunctional assemblies capable of performing under extreme conditions renders micro-joining essential. The development of microthrusters, infrared sensors, micro-gyros, and CubeSat components is facilitated by this support. Methods such as laser diffusion bonding and anodic joining are employed for ceramic-metal interfaces designed to endure vibration, vacuum, and radiation.

5.12.3 Future Scope and Technological Integration

In the future, new technologies are anticipated to transform the boundaries and potential of micro-joining. Laser micro-joining, especially with ultrafast femtosecond and picosecond lasers, facilitates the cold joining of dissimilar materials while ensuring minimal heat diffusion. These lasers facilitate the precise joining of glass to metal, transparent ceramics to semiconductors, and polymer-metal hybrids with unparalleled accuracy. This holds particular significance in the fields of photonics, optical MEMS, and biomedical packaging.

The combination of AI and digital twin platforms is transforming process control in micro-joining. Utilizing real-time sensor data allows for the prediction of bonding outcomes, the identification of process anomalies, and the adaptive tuning of parameters through AI. Digital twins facilitate the simulation of thermal gradients, stress distributions, and interfacial chemistry before physical joining, which helps to minimize errors and shorten process development cycles.

Nanojoining is an area experiencing significant growth in interest. The emergence of nanoscale electronics, quantum computing, and NEMS highlights the critical importance of joining at submicron levels. Investigations are underway into EBID, FIB welding, and van der Waals-based joining techniques to facilitate the seamless bonding of nanowires, nanotubes, and thin films.

Flexible and stretchable electronics, such as wearable sensors, smart textiles, and soft robotics, require micro-joining techniques that preserve mechanical integrity while bending or stretching. Advancements like liquid metal joining, ACFs, and photonic sintering of inks offer dependable interconnects that are not only functional but also adaptable.

5.12.4 Forward-Looking Applications and Research Directions

Micro-joining is set to emerge as a crucial technology within the realms of AM and 3D printing. Hybrid assemblies that include printed microelectronic components, embedded sensors, and functional circuits will depend on micro-joining to form integrated devices. The integration of micro-joining techniques with direct-write technologies, including aerosol jet and inkjet printing, will facilitate continuous and automated hybrid manufacturing processes.

Quantum devices and superconducting systems, which require cryogenic compatibility and ultra-clean interfaces, are poised to gain significant advantages. Micro-joining techniques such as low-temperature eutectic bonding or ALD-assisted joining maintain the integrity of quantum circuits while offering mechanical strength, all without adding thermal or electrical noise.

Innovative developments are unfolding in the area of adaptable and self-repairing joints. Drawing inspiration from biological systems, engineers are developing self-healing polymers, dynamic bonding agents, and reversible adhesives for application in mission-critical scenarios. These materials possess the ability to autonomously mend microcracks or adjust to new configurations while maintaining their functionality.

The drive toward sustainable manufacturing is promoting the advancement of biodegradable, energy-efficient joining techniques. This encompasses laser-releasable adhesives, water-soluble solders, and solvent-assisted bonding techniques tailored for environmentally friendly electronics and single-use biomedical devices.

Moreover, micro-joining is broadening its scope to include applications in extreme environments such as space, cryogenics, and areas with high radiation levels. Innovative materials and interlayers, including beryllium-copper and ceramic adhesives, are undergoing evaluation for their durability. These advancements guarantee that microsystems operate consistently in vacuum conditions, extreme cold, or nuclear environments.

5.12.5 Final Thoughts and Outlook

To sum up, micro-joining is no longer reduced to a "down-scaled" conventional welding or bonding, but it is the resulting state-of-the-art of a convergence of material science, microfabrication, mechanical engineering and digital (intelligence). The path is rife with technical challenges both subsurface—around fine-tuned control and optimum interfacial chemistry—but the potential applications are groundbreaking. The future of micro-joining is characterized by the ability to meet emerging needs: ultraminiaturization, multifunction, flexibility, sustainability, and intelligence. AI, ultrafast laser technology and new materials are enabling levels of precision and personalization at the micro- and nanoscale that have never been seen before.

Realizing this potential requires continuing to invest in development, standardization and inter-sectoral cooperation. The current lack of real-time process control, interfacial modeling and quality control must be overcome. It is crucial that industries work with academic and regulatory bodies to agree on metrics, validate methods and ensure interoperability. As we transition into an era of smart, autonomous and sustainable systems, micro-joining will be inscribed in the future of manufacturing. The contributions will cut through numerous fields—quantum computers on the one hand, perhaps wearable health on the other—and as the technology matures, it will serve as a mirror to the technological ambitions of our time.

REFERENCES

[1] Kutz, M. 2018. *Handbook of Materials Selection*. Wiley-Interscience.

[2] Kim, H.J., Lee, C. and Cho, Y. 2019. Dissimilar materials microjoining techniques in microelectronics: A review. *Materials*, 12(23), p. 3895.

[3] Abe, T., Yamaguchi, M. and Imai, H. 2001. Microjoining and nanojoining technologies for electronics applications. *Journal of Materials Processing Technology*, 113(1–3), pp. 51–56.

[4] Tummala, R.R. 2001. *Fundamentals of Microsystems Packaging*. McGraw-Hill.
[5] Rauscher, M., Plapper, P. and Stein, J. 2015. Micro joining technologies in medical device manufacturing. *Procedia CIRP*, 33, pp. 437–442.
[6] Oelgoetz, P. and Hilton, P.A. 2014. Microjoining in aerospace and defense manufacturing. In: Allen, D.M. (ed.), *The Principles of Laser Materials Processing*, pp. 457–478. Springer.
[7] Fan, Z., Zhang, Y. and Sun, L. 2020. Laser micro-welding for microjoining applications: Challenges and opportunities. *Optics & Laser Technology*, 129, p. 106278.
[8] Zhou, Y., Liao, Y. and He, X. 2022. Recent advances in sustainable microjoining technologies. *Journal of Manufacturing Processes*, 76, pp. 1035–1049.
[9] Zhou, Y. 2008. *Microjoining and Nanojoining*. Woodhead Publishing.
[10] Mian, A.J., et al. 2014. Laser transmission welding of polymers for microfluidics fabrication: A review. *Optics and Lasers in Engineering*, 55, pp. 8–18.
[11] Balle, F., Wagner, G. and Eifler, D. 2012. Ultrasonic metal welding of light metals to carbon fibre reinforced thermoplastic composites. *Materials Science and Engineering A*, 528(27), pp. 8287–8295.
[12] Roesner, H., Arguello, J.A. and Suganuma, K. 2001. Microjoining of ceramic and metal for high-temperature electronics applications. *Journal of Materials Processing Technology*, 115(3), pp. 394–398.
[13] Messler, R.W. 2004. *Principles of Welding: Processes, Physics, Chemistry, and Metallurgy*. Wiley-Interscience.
[14] Balle, F., et al. 2012. Micro laser welding of dissimilar materials in electronics. *Welding in the World*, 56(9–10), pp. 81–89.
[15] Katayama, S. 2005. *Laser Welding*. Springer.
[16] Tu, K.N. 2003. Recent advances on electromigration in very-large-scale-integration of interconnects. *Journal of Applied Physics*, 94(9), pp. 5451–5473.
[17] Zhou, Y. and Hu, A. 2013. *Microjoining and Nanojoining*. Woodhead Publishing.
[18] Steen, W.M. and Mazumder, J. 2010. *Laser Material Processing*, 4th edition. Springer.
[19] Kou, S. 2003. *Welding Metallurgy*, 2nd edition. Wiley-Interscience.
[20] Frear, D.R., Vianco, P.T. and Zender, G.L. 1997. Materials issues in microelectronics packaging. *MRS Bulletin*, 22(1), pp. 21–27.
[21] Suganuma, K. 2001. Advances in lead-free electronics soldering. *Current Opinion in Solid State and Materials Science*, 5(1), pp. 55–64.
[22] Lau, J.H. 1997. *Underfill Materials for Flip Chip Packaging*. Springer.
[23] Geiger, M., Vollertsen, F. and Schmidt, M. 2001. Laser beam micro welding—process and applications. *CIRP Annals*, 50(1), pp. 177–180.
[24] Roman, H.T. 1990. Thermal management issues in microelectronics. *IEEE Transactions on Components, Hybrids, and Manufacturing Technology*, 13(4), pp. 1040–1045.
[25] Kim, S. and Lee, J.H. 2012. Temperature management in electronics packaging. In: *Thermal and Thermomechanical Modeling of MEMS Packages*, pp. 19–34. Springer.
[26] Mahendran, R. and Yang, G. 2017. Challenges in thermal modeling of microelectronics packaging. *Microelectronics Reliability*, 76–77, pp. 493–504.
[27] Koenig, M., Vollertsen, F. and Schmidt, M. 2007. Microscale joining: State-of-the-art and future challenges. *Journal of Materials Processing Technology*, 189(1–3), pp. 335–343.
[28] Liewald, M. and Kinsey, B. 2013. Challenges in micro joining and forming processes. *CIRP Journal of Manufacturing Science and Technology*, 6(1), pp. 1–8.
[29] Kim, C. and Jeong, H. 2015. A review of micromanufacturing techniques and their applications. *International Journal of Precision Engineering and Manufacturing*, 16(5), pp. 943–957.

[30] Harman, G. 2010. *Wire Bonding in Microelectronics: Materials, Processes, Reliability, and Yield*. McGraw-Hill Education.
[31] Tsau, C.H., et al. 2004. Cleaning and activation of bonding surfaces for wafer-level bonding. *Journal of The Electrochemical Society*, 151(6), pp. G398–G403.
[32] Lee, J.N., Park, C. and Whitesides, G.M. 2003. Solvent compatibility of poly(dimethylsiloxane)-based microfluidic devices. *Analytical Chemistry*, 75(23), pp. 6544–6554.
[33] Madou, M.J. 2011. *Fundamentals of Microfabrication and Nanotechnology*, Vol. III. CRC Press.
[34] Bean, K.E. 1992. Silicon-based sensor arrays with microfabricated interconnects. *IEEE Transactions on Electron Devices*, 39(3), pp. 613–620.
[35] Vollertsen, F., Thomy, C. and Kallweit, T. 2008. Micro welding of thin wires and foils. *Welding in the World*, 52(5–6), pp. R96–R104.
[36] Liao, Y., Zhou, Y. and He, X. 2015. Low-temperature joining of ultra-thin wires for microelectronic interconnects. *Journal of Materials Processing Technology*, 221, pp. 54–62.
[37] Yang, G., et al. 2017. Challenges and solutions in joining of ultrathin flexible electronics. *Advanced Materials Interfaces*, 4(15), pp. 1700274.
[38] Eifler, M. and Kropp, M. 2014. Challenges in ultrasonic wire bonding and approaches for real-time quality monitoring. *Microelectronics Reliability*, 54(9–10), pp. 1962–1968.
[39] Gao, W. and Song, Z. 2018. Nondestructive evaluation methods for microelectronic packaging: State of the art. *IEEE Transactions on Components, Packaging and Manufacturing Technology*, 8(10), pp. 1795–1806.
[40] Kolar, M., et al. 2017. Inline inspection and AI for adaptive bonding process optimization. *Journal of Manufacturing Systems*, 44, pp. 190–199.
[41] Laurila, T., Vuorinen, V. and Kivilahti, J.K. 2005. Interfacial reactions between lead-free solders and common base materials. *Materials Science and Engineering R: Reports*, 49(1–2), pp. 1–60.
[42] Kim, K.S. and Tu, K.N. 1996. Kinetics of spontaneous solder reaction with Cu. *Journal of Applied Physics*, 80(10), pp. 5410–5415.
[43] Tan, C.M. and Lim, C.T. 2006. Reliability assessment and failure analysis of electrical interconnects. *Microelectronics Reliability*, 46(12), pp. 1929–1937.
[44] Lee, J., et al. 2020. Advanced packaging and interconnection techniques for next-generation electronics. *Micromachines*, 11(7), p. 623.
[45] Suganuma, K. 2014. Advances in lead-free electronics soldering. *Current Opinion in Solid State and Materials Science*, 18(2), pp. 55–60.
[46] Lau, J.H. 2017. *Fan-Out Wafer-Level Packaging*. Springer.
[47] De Falco, A., et al. 2019. Advanced microjoining techniques for biomedical applications. *Journal of Materials Science: Materials in Medicine*, 30(4), p. 39.
[48] Guo, S.Z., et al. 2016. Bioinspired 3D structures for microscale biomedical devices. *Advanced Materials*, 28(35), pp. 7163–7169.
[49] Zhang, X. and Mei, Z. 2014. Sustainable microjoining: Green approaches and prospects. *Materials Science and Engineering B*, 185, pp. 42–49.
[50] Rao, R.R., et al. 2015. Environmental considerations in advanced manufacturing processes: Case study of printed electronics. *Procedia Manufacturing*, 1, pp. 436–445.
[51] Kim, D.H., et al. 2011. Epidermal electronics. *Science*, 333(6044), pp. 838–843.
[52] Lin, L., et al. 2017. Microjoining for aerospace applications: Current status and future perspectives. *Journal of Materials Processing Technology*, 247, pp. 1–15.
[53] Ramesh, R. and Tzeng, C. 2016. Advanced joining techniques for aerospace and defense applications. *Materials Performance and Characterization*, 5(3), pp. 425–440.

[54] Xu, S., et al. 2014. Soft microfluidic assemblies of sensors, circuits, and radios for the skin. *Science*, 344(6179), pp. 70–74.
[55] Kim, J., et al. 2022. Emerging materials and processes for stretchable and wearable electronics. *Advanced Materials*, 34(10), p. 2106626.
[56] Weber, R., et al. 2018. Femtosecond laser welding of glasses. *Laser & Photonics Reviews*, 12(4), p. 1700273.
[57] Katayama, S. 2013. *Handbook of Laser Welding Technologies*. Woodhead Publishing.
[58] Zhang, X., et al. 2015. Nanojoining: Principles, technologies and applications. *Microelectronics Reliability*, 55(9–10), pp. 1666–1676.
[59] Utke, I., Hoffmann, P. and Melngailis, J. 2008. Gas-assisted focused electron beam and ion beam processing and fabrication. *Journal of Vacuum Science & Technology B*, 26(4), pp. 1197–1276.
[60] Kim, J.S., et al. 2010. Self-assembled nano-joining using metal nanoparticles for flexible electronics. *ACS Nano*, 4(2), pp. 456–464.
[61] Wang, X., et al. 2022. Artificial intelligence for smart manufacturing: Micro joining applications and challenges. *Journal of Manufacturing Systems*, 62, pp. 705–720.
[62] Tao, F., et al. 2019. Digital twin driven smart manufacturing: Connotation, reference model, applications and research issues. *Robotics and Computer-Integrated Manufacturing*, 61, p. 101837.
[63] Liu, Z. and Li, Y. 2021. AI-driven process optimization in precision micro joining. *Materials Today: Proceedings*, 38, pp. 1281–1287.
[64] Zhao, H., Wang, Y., Liu, Y., et al. 2020. Application of deep learning in welding defect detection: A review. *Journal of Manufacturing Processes*, 56, pp. 1082–1097.
[65] Huang, Y., Zhang, Z., Sun, Z., et al. 2021. Artificial intelligence in welding: A review on machine learning and deep learning approaches. *Materials Today Communications*, 28, p. 102645.
[66] Bovatsek, J., Patel, R., Alghannam, A., et al. 2019. Femtosecond laser micromachining for microelectronics packaging applications. *Journal of Laser Applications*, 31(2), p. 022005.
[67] Mahrle, A. and Beyer, E. 2019. Hybrid laser welding—A literature review. *Journal of Laser Applications*, 31(4), p. 041202.
[68] Zhou, Y., Mahalingam, R. and Zhang, J. 2020. *Microjoining and Nanojoining*. Woodhead Publishing.
[69] Chen, Y., Liao, Y. and Liu, X. 2021. Energy-efficient solid-state microjoining: A review. *Micromachines*, 12(6), pp. 635.
[70] Sun, C., Watanabe, S. and Fatikow, S. 2022. Micro joining and hybrid additive manufacturing: Challenges and prospects. *Additive Manufacturing*, 53, p. 102703.
[71] Gibson, I., Rosen, D. and Stucker, B. 2021. *Additive Manufacturing Technologies: 3D Printing, Rapid Prototyping, and Direct Digital Manufacturing*. Springer.
[72] Guo, Y., Li, H., Zhang, L., et al. 2020. Nanojoining for nanoelectronics: Technologies, challenges, and future directions. *Nano Today*, 33, p. 100872.
[73] Zhang, Y., Sun, L. and Zhang, H. 2021. Micro- and nanojoining of advanced composite materials: Progress and perspectives. *Advanced Materials Interfaces*, 8(21), p. 2100883.
[74] Kim, D.H., Ghaffari, R., Lu, N. and Rogers, J.A. 2019. Flexible and stretchable electronics for biointegrated devices. *Annual Review of Biomedical Engineering*, 21, pp. 113–142.
[75] Liu, Y., Pharr, M. and Salvatore, G.A. 2021. Lab-on-skin: A review of flexible and stretchable electronics for wearable health monitoring. *ACS Nano*, 15(4), pp. 6051–6070.

[76] Wang, Y., Zhao, D., Xu, X., et al. 2022. Advances in low-temperature bonding for MEMS and quantum device packaging. *Journal of Micromechanics and Microengineering*, 32(5), p. 053002.
[77] Nguyen, L.T. 2021. Packaging of quantum and MEMS devices: Challenges and opportunities. *Microelectronics Reliability*, 122, p. 114143.
[78] Toohey, K.S., Sottos, N.R., Lewis, J.A., et al. 2007. Self-healing materials with microvascular networks. *Nature Materials*, 6(8), pp. 581–585.
[79] Wang, C., Xu, J., Zhang, X., et al. 2020. Dynamic and reconfigurable electronics: Current status and future directions. *Advanced Functional Materials*, 30(33), p. 2001059.
[80] Feng, Y., Song, W., Zhang, G., et al. 2019. In situ sensing technologies in welding and joining: A review. *Journal of Manufacturing Systems*, 52, pp. 1–17.
[81] Wang, H., Xu, Y., Zhang, X., et al. 2021. Real-time monitoring and control in micro joining processes using advanced sensors. *Sensors and Actuators A: Physical*, 323, p. 112638.
[82] Kwon, H., Song, S., Jeong, Y., et al. 2021. Micro joining technologies for harsh environments: Challenges and developments. *Journal of Materials Science: Materials in Electronics*, 32, pp. 21378–21393.
[83] Lee, J., Kang, S., Lee, K., et al. 2020. Cryogenic and vacuum-compatible micro joining techniques for space applications. *Materials & Design*, 195, p. 109033.
[84] Thompson, M.K. & Moroni, G. 2023. Artificial intelligence in advanced manufacturing systems. *CIRP Annals*, 72(2), pp. 1–24.
[85] Wang, J., Ma, Y., Zhang, L., et al. 2023. Deep learning for smart manufacturing: Methods and applications. *Journal of Manufacturing Systems*, 68, pp. 1–15.
[86] Zhang, C. & Chen, Y. 2022. Intelligent laser welding based on machine learning and real-time monitoring. *Optics and Laser Technology*, 156, p. 108456.
[87] Zhang, W. & Li, R. 2022. Reinforcement learning for adaptive control of micro-joining processes. *Engineering Applications of Artificial Intelligence*, 116, p. 105387.
[88] Gonzalez, F. & Liu, Y. 2022. Computer vision and deep learning for real-time weld quality assessment. *Robotics and Computer-Integrated Manufacturing*, 78, p. 102401.
[89] Schmidt, M. & Richter, K. 2023. AI-based process monitoring and control in resistance micro-welding. *Journal of Manufacturing Processes*, 85, pp. 1–12.
[90] Kim, H. & Park, S. 2023. Digital twin-driven quality control in micro-electronics packaging. *IEEE Transactions on Components, Packaging and Manufacturing Technology*, 13(4), pp. 567–578.
[91] Li, X., Wang, Z. & Zhang, H. 2022. Machine learning in micro-joining: A comprehensive review. *Journal of Materials Processing Technology*, 309, p. 117732.

6 Typical Case Studies and Applications

6.1 INTRODUCTION

One of the most fundamental manufacturing processes is joining, by which it is feasible to bring together disparate components into a functional assembly. Joining processes can generally be classified into two types: micro-joining and macro-joining, based on the application size, precision required, and material properties. To achieve the necessary structural, electrical, or operational efficiency, every segment has specific technical challenges and requires the utilization of specialized technologies.

Micro-joining is the assembly and creation of parts with usual dimensions ranging from a few micrometers to a few millimeters. Optoelectronics, microelectronics, medical implants, sensors, and microelectromechanical systems (MEMS) are some of the industries that make use of this scale by default. The issues at this scale are retaining electrical conductivity, avoiding microstructural damage, reducing thermal distortion, and accurately regulating heat. Techniques including diffusion bonding, resistance micro-welding, ultrasonic welding, thermosonic wire bonding,[1] and laser micro-welding are frequently used. For instance, thermosonic bonding is used to link copper or gold wires smaller than 25 μm in diameter to integrated circuit pads in semiconductor packaging, ensuring low electrical resistance and mechanical dependability.

The term "macro-joining" describes procedures used for large-scale assembly, usually ranging from millimeters to meters, where the focus is on fatigue resistance, joint strength, load-bearing capacity, and production efficiency. It is essential to sectors including heavy engineering, pipeline construction, aerospace, shipbuilding, automotive production, and infrastructure for renewable energy. Resistance spot welding, friction stir welding, submerged arc welding, thermite welding, and arc welding procedures (gas tungsten arc welding, shielded metal arc welding, and gas metal arc welding) are examples of common techniques. These procedures guarantee long service life and structural integrity while managing thicker portions and higher heat input.

There are a number of considerations when deciding between micro- and macro-joining techniques:

- Dimensional tolerances and precision requirements
- Material qualities, such as thermal conductivity, melting point, and coefficient of thermal expansion
- Type of joint and loading circumstances
- Cost and production volume limitations
- Service and environmental circumstances (such as temperature cycling, vibration, and corrosion)

DOI: 10.1201/9781003502395-6

A practical understanding of the way these elements affect process selection, tooling design, quality control techniques, and overall manufacturing strategy can be gained by studying real-world case studies and industrial applications of both micro- and macro-joining. Additionally, they illustrate how technical advancements like automation, laser processing, and hybrid joining are allowing for more dependable, effective, and economical assembly by bridging the gap between the micro- and macro-domains.

6.2 CASE STUDY 1: MICROELECTRONIC CHIP BONDING USING THERMOSONIC BALL BONDING

6.2.1 Introduction to Microelectronic Chip Bonding

Integrated circuits (ICs) are at the heart of practically every device in the modern electronics industry, from laptops and smartphones to medical implants and aerospace navigation systems. Transistors and interconnects—two minuscule components found in silicon wafers—are the center of IC manufacture, yet they cannot operate independently. Chip packaging is the process of connecting a manufactured chip to the external environment both mechanically and electrically. Chip bonding, also known as die connectivity, is a crucial step in this process because it creates the electrical and physical connection between the semiconductor die and the printed circuit board (PCB) or package leads. Among the different bonding techniques, thermosonic ball bonding is well-known for producing fine-pitch, extremely dependable connections that are appropriate for high-performance applications. It is particularly common in fields where signal integrity, dependability, and downsizing are crucial, such as consumer electronics, automotive electronics, aircraft systems, and biomedical devices.

6.2.2 Overview of Thermosonic Ball Bonding

Thermosonic bonding is a micro-joining technique that creates a solid-state weld between a bonding wire and the bonding pad on the semiconductor device or substrate by combining thermal energy, ultrasonic vibration, and mechanical pressure. A spherical ball at the wire's tip is used in ball bonding to create the initial connection. The ball is then pressed into the bond pad. The second bond, which is frequently referred to as a stitch or wedge bond, is then produced at the matching pad on the package lead or substrate after a wire loop has been formed. To create the metallurgical connection, this approach uses controlled deformation and interatomic diffusion at the interface rather than melting the bulk materials being joined. Although copper and aluminum are also utilized in some applications, gold has long been the most widely used wire material because of its ductility, stability in bonding, and resistance to corrosion.[2]

6.2.3 Process Steps

The creation of a known free air ball at the end of the bonding wire initiates the thermosonic ball bonding process. A tiny portion of the wire emerges from the tip after being run through a ceramic capillary tool. A high-voltage electronic flame-off (EFO) discharge melts this part, and the molten metal is drawn into a nearly flawless sphere by surface tension. By modifying the EFO parameters and the wire diameter, the size of this ball may be precisely adjusted.[3]

The bonding tool positions the produced ball directly over the chip's bond pad. The first bond is formed by heating the chip or bonding stage, applying ultrasonic vibrations to the tool, and pressing the ball against the pad under a predetermined bond force. Heat and ultrasonic scrubbing work together to remove oxide layers and surface impurities, enabling atomic-level metal-to-metal contact. The ball slightly deforms under pressure, expanding the contact area and starting the solid-state diffusion between the pad metallization and the ball.

The bonding tool rises and proceeds to the second bond's site after the first bond is formed, paying out the wire to form a loop. Because it influences the connection's mechanical dependability during handling, encapsulation, and thermal cycling, the loop design is crucial. Without forming a ball, the wire is pressed against the metallized pad on the substrate or package lead to create the second bond. We refer to this as the wedge bond or stitch. After applying moderate force and ultrasonic energy to guarantee a strong weld, the bonding tool is moved to the side and upward, clamping the wire and breaking it off. Every time a wire connection is needed, the procedure is repeated.[4]

6.2.4 Metallurgical Principles

Solid-state welding principles, which establish the bonding interface without melting the bulk material, are the foundation of the thermosonic ball bonding process. By causing microscopic slip and relative motion between the mating surfaces, the delivered ultrasonic energy breaks down impurities and oxide layers.[5] Heat simultaneously increases atomic mobility and decreases the materials' yield strength, while the applied force guarantees close surface contact. Intermetallic compounds like $AuAl_2$ and Au_2Al are created at the interface when gold wire is connected to aluminum pads.[6] Although these compounds have strong mechanical strength and electrical conductivity, their thickness needs to be regulated since too much growth over time can cause brittle fracture or void formation, a phenomenon called Kirkendall voiding.[7]

6.2.5 Equipment and Tooling

A precision wire bonding machine is used to carry out the thermosonic bonding procedure. The capillary tool, the wire clamping mechanism, and the bond head—which houses the ultrasonic transducer are essential parts. The capillary guides the wire and transfers ultrasonic energy to the point of bonding; it is typically made of a durable ceramic, such as alumina-titanium carbide.[8] The thermal component of the process is provided by the work holder, which also holds and heats the die or package. Additionally, the machine contains an EFO unit to precisely control the diameter of free air balls and an optical vision system to align the bonding tool with the tiny bond pads. The instruments can make over ten bonds per second and offer high repeatability and throughput.[9]

6.2.6 Industrial Applications

Thermosonic ball bonding shows up a lot in various industries. They use it for packaging microelectronic devices. In the smartphone world, it links up processors and memory chips. It also handles other dense components to their bases. Pretty much, the precision and consistency of this bonding really matter. These gadgets often pack in thousands of bond pads. The spacing can get down to just 35 or 40 micrometers. Over in the auto sector, folks rely on it for safety setups and sensors. It goes into control modules too. Dependability counts big time there. This is particularly crucial in light of the extreme temperatures these modules encounter. Fluctuations and mechanical vibration are essential. Gold ball bonding is also necessary for the biocompatibility and long-term stability of medical equipment like pacemakers and hearing aids inside the human body. Thermosonic bonding offers reliable performance over extended mission lifetimes in aerospace and defense electronics, where devices must function in harsh conditions.

6.2.7 Advantages of Thermosonic Ball Bonding

The capacity of thermosonic ball bonding to create sturdy, dependable connections at comparatively lower temperatures than thermocompression bonding, which lowers the possibility of heat damage to delicate semiconductor devices, is one of its main advantages. The process's ultrasonic component allows it to withstand tiny oxide coatings, allowing bonding to aluminum pads without requiring a lot of pre-cleaning. The procedure supports high-volume manufacturing since it is quick and highly repeatable. Electronic goods can continue to get smaller thanks to their fine-pitch capability, which enables designers to fit more I/O connections into smaller package sizes. The resultant bonds satisfy functional and durability criteria with their strong mechanical strength and low electrical resistance.

6.2.8 Limitations and Challenges

Although thermosonic ball bonding has many benefits, there are drawbacks as well. Although gold wire has excellent bonding properties, it is expensive, and changes in the price of gold can have a big impact on the cost of manufacturing high-volume goods. Over time, bond failure may result from the development of brittle intermetallic compounds at the gold-aluminum contact, particularly in situations involving high storage or cycling temperatures. Although copper wire is more affordable and has superior electrical conductivity, it oxidizes more easily and needs regulated bonding environments. Maintaining the loop geometry presents another difficulty because deviations may result in wire sweep during plastic encapsulation, which could result in open connections or short circuits.[10]

6.2.9 Quality Control and Reliability Testing

Both in-process monitoring and post-bond inspection are necessary to guarantee the quality of thermosonic ball bonding. Optical microscopy frequently examines the shape, location, and deformation of the bonded balls, as well as the height and consistency of the loops. The creation of intermetallic layers and the bond interface can be thoroughly examined using scanning electron microscopy. Hidden flaws like broken wires, heel cracks, or shorts between wires can be found using X-ray imaging. Ball shear tests, which gauge the lateral force needed to separate the ball from the pad, and wire pull tests, which evaluate the strength of the first and second bonds, are examples of mechanical testing techniques. To guarantee long-term dependability under operational settings, environmental stress tests such as thermal cycling, high-temperature storage, and autoclave moisture exposure are also carried out.

6.2.10 Case Example: Thermosonic Bonding in a Smartphone System-on-Chip (SoC)

Packaging a 7-nanometer SoC with more than 1,500 I/O pads at a pitch of only 40 micrometers was a problem for a top smartphone manufacturer. The gadget needed minimal signal delay and low electrical resistance connections to withstand drop tests and function dependably in temperatures ranging from –20°C to 85°C. The business evaluated several connecting methods before deciding on gold thermosonic ball bonding because of its track record in high-reliability, fine-pitch applications.[11]

To match the pad metallurgy, 18 micrometers of gold wire with a stage temperature of 170°C and ultrasonic energy optimized at 96 kHz were used. High-volume production was made possible by the bonding process's achievement of a cycle time of 85 milliseconds per wire. The ball shear strength continuously exceeded the minimum requirement by 15%, and the first-pass yield was above 99.8%. After 1,000 heat cycles, reliability tests revealed no bond failures, confirming the process selection. The manufacturer defended the choice by pointing to the combination of high yield, demonstrated dependability, and performance stability, even though gold was pricier than copper.[12]

6.2.11 Future Trends in Thermosonic Bonding

To save costs without sacrificing performance, the industry is looking more and more into copper and palladium-coated copper as substitutes for pure gold wire. Artificial intelligence is being included in bonding equipment advancements for adaptive loop management, guaranteeing consistent geometry even in the face of environmental changes or machine drift. To support flexible electronics and devices with thermally sensitive materials, low-temperature bonding techniques are being developed. Furthermore, thermosonic bonding is evolving to meet the demands of sophisticated three-dimensional packaging, such as stacked die configurations and through-silicon-via interposers, where interconnection density and dependability are even more crucial.[13]

6.2.12 Summary

One of the most crucial micro-joining techniques in the production of microelectronics is still thermosonic ball bonding. For semiconductor packaging in a variety of industries, its capacity to generate high-reliability, fine-pitch connections with regulated thermal impact makes it essential. The thorough analysis of its use in a smartphone SoC demonstrates its advantages in high-volume, performance-critical production settings. Thermosonic ball bonding remains the industry standard in applications where process maturity, accuracy, and long-term dependability are more important than material cost considerations, notwithstanding the emergence of alternative bonding technologies. With continuous advancements in materials, machinery, and process control, thermosonic bonding will continue to play a crucial part in electronics' continued evolution toward greater integration and miniaturization.

6.3 CASE STUDY 2: MEMS SENSOR ASSEMBLY USING LASER MICRO-WELDING

6.3.1 Introduction to MEMS and Sensor Integration

An intricate family of devices at the micro-scale, termed microelectromechanical systems (MEMS), is made up of electronic circuitry and mechanical build. MEMS structures, such as accelerometers and pressure sensors, have moving pieces that react to stimuli and range in size from about 20 micrometers to some millimeters. Their precision in integrating ancillary electronics, such as signal processing circuits, into compact packages is key to their reliability and performance.[14] The selection of the connecting technique becomes increasingly important when these devices miniaturize and are used in demanding applications such as industrial sensing, medical instrumentation, and aerospace. Traditional bonding techniques could lead to mechanical stress, undue heat introduction, or a loss of hermeticity necessary for sensor operation. Due to the delicate nature of MEMS sensor fabrication, laser micro-welding emerges as a powerful technology here with its low-heat joining accuracy.

6.3.2 Technical Overview of Laser Micro-Welding

Laser micro-welding is one of the very advanced forms of welding aimed at minute and fragile parts. The process generates localized heat that melts and joins materials in milliseconds without needing to employ fillers or adhesives by focusing a high-energy laser beam on closely spaced locations. One of the key features of MEMS packages is the minimum heat-affected zone, which ensures that the surrounding fragile structures are not damaged. With spot diameters often below 50 micrometers, continuous-wave fiber lasers, as well as pulsed regimes (e.g., Nd:YAG lasers), can be employed. High-precision weld location and quick processing within highly limited designs are made possible by the laser beam's accuracy and the possible use of galvanometer steering heads. Notably, laser micro-welding allows for the connecting of dissimilar and similar materials, facilitating hermetic sealing, which is essential for the longevity of sensors. The deliberate application of inert atmospheres enhances the quality of the weld and guards against contamination or oxidation throughout the procedure.[15]

6.3.3 Process Steps in MEMS Sensor Assembly

A meticulously planned procedure is required to assemble MEMS pressure or accelerometer sensors with their microcircuits using laser micro-welding. MEMS components and electronic microcircuit segments are first precisely aligned, frequently with the use of high-resolution stages that can put objects with submicron accuracy. The laser is then used to make weld seams at joints such as connecting frames, lids, or structural supports after being aligned in a controlled environment, usually inside a glovebox filled with inert gases like argon. A strong weld without overexposure is achieved by adjusting the laser's power, pulse duration, spot size, and repeat frequency to minimize melting. Pulsed Nd:YAG lasers are frequently employed for spot or seam welding of MEMS enclosures because they can generate 1–5 joules per pulse at peak power of up to several kilowatts, providing joint integrity and misalignment tolerance. Using galvanometer scan heads, the welding motion may be precisely and swiftly adjusted to cover small areas, guaranteeing repeatable seams across several devices.[16]

6.3.4 Metallurgical and Structural Considerations

At the micro-scale, the joining region's microstructure defines both mechanical strength and the functional longevity of the device. In laser micro-welding, the rapid melting and resolidification foster a clean metallurgical bond often free from contaminants or filler metals. The extremely localized heating significantly limits diffusion-driven degradation, preserving nearby sensitive electronics and maintaining structural tolerances. For hermetic sealing, especially of pressure sensors, ensuring a continuous, void-free weld seam is essential to prevent gas or moisture ingress. The integrity of this weld depends on careful control of thermal gradients; using a pulsed laser with short duty cycles helps minimize thermal stresses while achieving adequate bond penetration.

6.3.5 Equipment and Tooling

An Nd:YAG pulsed laser system with fiber-optic delivery for flexibility and stability is a typical laser micro-welding station in MEMS assembly. Galvanometer mirrors may be employed to steer or shape the beam to facilitate rapid micrometer-precise motion without issues inherent with the motion of moving parts. The MEMS component stage, often associated with vision systems for automated positioning, needs to ensure vibration-damped precision alignment along the X, Y, and Z axes. Environments need inert gas shielding and, in sensitive situations, vacuum packing conditions to avoid oxidation or moisture contamination. A Class I laser-safe enclosure with interlocks and operator safety might contain the entire procedure. This advanced tooling enables high-throughput, repeatable processing while maintaining the precise tolerances necessary for MEMS packing.

6.3.6 Industrial Applications of Laser Welding in MEMS Packaging

Laser micro-welding is becoming more and more acknowledged as the best technique for integrating MEMS sensors in many industries. Tightly regulated laser welds provide biocompatibility and hermetic sealing, which are advantageous for implantable pressure sensors in the medical area, such as those used for intracranial pressure monitoring. Accelerometers for guiding systems in aerospace and defense need strong, thermally stable packaging that can tolerate high vibration and challenging climatic conditions; laser sealing provides long-lasting integrity with little mass addition. Similar to this, laser welding is used in automotive applications, such as tire-pressure monitoring sensors, to guarantee hermeticity and long-term endurance under vibration, temperature cycling, and contaminant exposure.[17]

6.3.7 Advantages of Laser Micro-Welding in MEMS Assembly

Laser micro-welding is especially well suited for MEMS sensor construction because it offers several technological benefits. The ability to transmit energy in a highly localized manner prevents harm to adjacent microelectronic components, maintaining the sensitivity and calibration of accelerometers or pressure sensors. Since there are frequently no fillers used, any pollutants and chemical reactions that can affect sensor performance are removed. The smooth, tiny weld joints enable high-density integration in compact packaging. Compact, multicomponent stacks that are necessary for integrating MEMS sensors with their signal-conditioning circuits and packaging lids are also supported by the exact control over weld placement and geometry. Laser pulse speeds and automation capabilities enable high manufacturing throughput, thereby increasing production yield and reducing device costs.

6.3.8 Limitations and Challenges

Laser micro-welding has drawbacks despite its advantages. Weak or irregular welds might result from fit-up tolerance, which is a tight misalignment that goes beyond a few microns. Upstream packing tolerances must be met in order to provide sufficient overlap and surface contact in the weld joint design. Surface finishes and material reflectivity, such as those of polished metals, can alter energy absorption and require modified laser parameters. Heat control is essential because thin MEMS structures can get stressed or distorted by even tiny temperature gradients. Laser systems and enclosures might have significant capital costs, and process development for novel sensor designs might necessitate a great deal of parameter tuning. Although it is possible, automation requires spending money on environmental control infrastructure, fixtures, and vision systems.

6.3.9 Case Example: Laser Micro-Welding in a MEMS Pressure-Accelerometer Hybrid Module

In a compelling industrial example, a manufacturer of inertial measurement units (IMUs) embarked on developing a compact hybrid module combining a MEMS pressure sensor with a MEMS accelerometer and micro-circuitry for drone altitude and motion control. The goal was to minimize package volume while ensuring hermeticity, mechanical strength, and signal integrity under high-vibration conditions typical in UAV operations[18]

The assembly process began with precise placement of the pressure sensor die onto a metallized frame, followed by bonding to the accelerometer die and PCB-level circuitry. Given the delicate diaphragm on the pressure sensor and fragile cantilevers of the accelerometer, traditional soldering or adhesive bonding risked damage or parasitic stiffness changes. Laser micro-welding was chosen to join the stainless-steel cover and frame that encapsulated the sensors.

Using a pulsed Nd:YAG laser system equipped with a galvanometer scanning head, the team programmed a weld seam around the cover's perimeter. The laser was configured for 2 joules per pulse with a peak power of 1 kW, delivering spot sizes of about 100 micrometers. Beam positioning was controlled to within ±5 micrometers. A nitrogen atmosphere prevented oxidation during welding.

The process yielded a continuous hermetic seam roughly 150 micrometers wide. Post-weld helium leak testing confirmed leak rates well below 1×10^{-9} atm·cm^3/s. Vibration testing at 20 g across a 20–2000 Hz band induced no degradation in signal output. Thermal cycling between –40°C and +85°C across 500 cycles produced negligible drift in sensor sensitivity (<0.1%). The localized heat input preserved the MEMS structures, with no mechanical alteration observed under scanning electron microscopy (SEM) inspection.

Despite high initial setup costs, the process was highly repeatable, achieving yields above 99% in production. The rationale for adopting laser micro-welding was rooted in achieving a compact, robust, and hermetically sealed module without compromising sensor performance. The added manufacturing complexity was justified by the improved reliability and scaled in high-volume drone systems.

6.3.10 Summary

A convincing method for combining MEMS accelerometer and pressure sensors with microcircuits in contemporary, miniature applications is laser micro-welding. Hermetic sealing, excellent mechanical integrity, and high yield in small packages are made possible by its precise, localized heat input, which maintains MEMS performance. The technology stands out as a dependable, scalable, and resilient micro-joining solution in sophisticated sensor assembly domains from aerospace to medicinal devices, even though deployment necessitates careful design, tooling investment, and parameter management.

6.4 CASE STUDY 3: MEDICAL DEVICE MICRO-ASSEMBLY USING RESISTANCE MICRO-WELDING FOR PT–IR WIRES IN PACEMAKERS AND NEUROSTIMULATORS

6.4.1 Introduction to Medical Device Micro-Assembly

Pacemakers and neurostimulators are just two instances of medical implantable devices that rely on extremely precise and durable connections between electrode structures and thin platinum-iridium (Pt–Ir) wires. For maintaining electrical function and biocompatibility, the micro-assemblies have to withstand decades-long periods of mechanical and physiological stresses. One critical method that has emerged to meet such stringent requirements is resistance micro-welding. To create specific metallic bonds with minimal thermal diffusion and maintain delicate components, controlled electrical resistance heating under pressure is employed in this method. Its excellent strength, biocompatibility, and corrosion resistance make it an ideal choice for joining materials such as Pt–Ir. Resistance micro-welding offers precision, repeatability, and acceptance in terms of very strict medical standards, making it better than soldering or laser methods in uses wherein the connection demands delicate electrode contacts or feedthroughs.[19]

6.4.2 Technical Overview of Resistance Micro-Welding

A high-current, precisely controlled low-voltage pulse is delivered between electrodes, clamping the parts together to allow for resistance micro-welding. The micro-scale contact surface resistance creates intense localized heat, softening or melting material enough to form a bond. Pressure from the electrodes ensures proper contact and avoids contamination or oxidation. This technique stands out in the medical assembly field due to its precise control over the energy supply, which significantly reduces unwanted heating effects. For thin wires such as Pt–Ir, whose electrical strength, resistance, and radiopacity qualify them for pacing and neurostimulation, the weld's small size—typically on the order of tens of micrometers—is ideal. Transistor welders and other sophisticated resistance micro-welding equipment allow us to regulate the current and form the pulses in real-time, enhancing the quality.

6.4.3 Process Steps in Pt–Ir Wire Welding

The process of attaching Pt–Ir wires by micro-welding begins with the micropositioning of the wire, electrode, or pad assembly under high magnification vision guidance. After aligning, carefully crafted welding electrodes press against the interface and send out a controlled current pulse that usually lasts a few milliseconds. Material fusion, or interdiffusion, results from the melting or plasticization of small, localized areas of the Pt–Ir wire and mating electrode during this pulse. The electrodes maintain pressure through cooling immediately following the current pulse to strengthen and perfect the connection. Contact resistance is essential because of the tiny scale: even short pulses can produce enough heat when there is a greater resistance at the contact. The engineer controls the following parameters: electrode force, surface preparation, current amplitude, and pulse duration, which determine bond morphology, strength, and dependability.[20]

6.4.4 Metallurgical and Biocompatibility Considerations

The metallurgical result and biological compatibility are non-negotiable in the micro-assembly of medical devices. Pt–Ir wires must fuse without the introduction of brittle or cytotoxic intermetallics because they are frequently connected to electrodes or pads made of alloys or noble metals. Resistance welding maintains material characteristics and interface integrity by allowing for minimum thermal diffusion. Microstructural investigation using SEM has revealed dense, defect-free welds without identifiable transition layers in tests welding Pt–Ir (e.g., Pt–10% Ir) to stainless steel or other alloys—indicating well-executed interdiffusion and fusion.

These structures maintain their stability in the face of mechanical and thermal stress without sacrificing biocompatibility, which is essential for devices that spend a lot of time within the human body.[21]

6.4.5 Equipment and Tooling

Pt–Ir micro-welded junctions require highly specialized micro-welding equipment with fine control over the parameters of the process, mechanical and electrical. Controlled pulses are generated by transistorized welders such as the Avio MCW-700/750 series in constant current, constant voltage, or constant power modes, the importance being to control the energy input at a very small scale. The welding heads are normally equipped with precision stages, vision systems, and micro-sized electrodes for alignment better than −10 μm. Some systems also feature online displacement, voltage, and current monitoring systems to detect anomalies early. A cleanroom environment or inert gas atmosphere is provided as added protection against oxidation or contamination. Medical assembly systems integrate software-controlled process documentation to ensure compliance with quality standards such as ISO 13485.

6.4.6 Industrial Applications in Medical Devices

The construction of pacemakers, implantable defibrillators, neural stimulators, and other active implantable medical devices (AIMDs) frequently uses resistance micro-welding. The technique's ability to weld thin Pt–Ir wires, often with a diameter of less than 50 μm, to electrodes or contact pads with mechanical and electrical dependability while adhering to biocompatibility criteria is valued by manufacturers. According to reports from businesses such as Cisteo MEDICAL, gold may be micro-welded to Pt–Ir, which is utilized in neurostimulator devices, in a matter of milliseconds without causing contamination or material distortion. When failure is intolerable, this precision makes it possible to produce class III medical implants.[22]

6.4.7 Advantages of Resistance Micro-Welding for Pt–Ir

For micro-assembly in medical implants, resistance micro-welding provides a number of benefits. It maintains sensitive structures and functional geometries by providing quick, repeatable, high-precision connections between small wires and electrodes without transferring too much heat. Because there are no fillers used, there is no chance of contamination. It provides exceptional joint strength, can be highly automated, and takes into account different materials (such as Pt–Ir and stainless steel). For high-volume production, welding times in the millisecond range maximize throughput. Last but not least, the micro-scale weld guarantees that the implant will continue to be small, strong, and biocompatible—a crucial combination for devices like pacemakers or neurological stimulators.

6.4.8 Limitations and Challenges

Micro-welding at such an extremely small scale presents challenges in establishing a resistance system. A misalignment of more than a few micrometers can lead to subpar or unsuccessful welding; therefore, tight fit tolerances are necessary. The Pt–Ir and electrode surfaces' cleanliness and surface finish have a significant impact on the contact resistance, which in turn affects the bond quality and heat distribution. Current pulses must be controlled to prevent overheating, which can destroy insulation or deteriorate material qualities. Pulse parameters need to be precisely adjusted, which frequently calls for design-of-experiments optimization and iterative testing. Furthermore, upfront expenditure is increased by the necessity for controlled settings and the cost of capital equipment; however, these are counterbalanced by regulatory compliance and dependability.

6.4.9 Case Example: Resistance Micro-Welding in a Miniature Pacemaker Lead Assembly

While participating in a program for the development of sophisticated medical devices, a group of individuals set out to create a next-generation ultra-miniature pacemaker that would be no bigger than a small coin and would be acceptable for use in pediatric applications or minimally invasive implantation. The design called for attaching electrode pads and microcoils inside a small titanium container to incredibly thin Pt–Ir alloy wires that were 15 μm in diameter. Traditional adhesive bonding or soldering ran the danger of leaving chemical residues or too much heat, which might damage the fragile insulation and reduce biocompatibility. The team chose resistance micro-welding due to its accuracy, low-heat diffusion, and reliable junctions.

The fine Pt–Ir wires were prepared for prototyping by carefully cutting and cleaning them to get rid of surface oxides. To avoid contamination, submicron robotic placement was used to line the electrode pads with the wires inside an ISO Class 7 cleanroom. The electrode pads were made on a flexible polyimide substrate that had been covered with a thin layer of noble metal. A transistor-based micro welder that could produce shaped pulses in constant current mode was the selected welding system.[23]

In the initial parameter investigation, electrode pressure was varied between 0.5 and 2 MPa, while current was varied between 200 and 600 A across pulse durations of 1 to 5 ms. SEM cross sections of sample joints showed that the Pt–Ir wire and pad fused without voids or undue distortion at approximately 400 A and 3 ms with 1 MPa of pressure. For the anticipated mechanical loads, the joint's shear strength of about 4 cN was more than enough. In the absence of clear brittle intermetallic layers, the contact displayed mutual diffusion.

Minor insulation breaches resulted from the degradation of insulation materials at higher current or longer pulses. Electrical resistance was too high, and the weld was weak or incomplete at lower energy values. The team determined the final parameters through pull tests, electrical testing, and repeated tuning based on SEM: 1.2 MPa pressure, 3 ms duration, and 450 A pulse. Ten wire-pad assemblies were positioned beneath the welding head every second by an automated vision-guided stage during production, with alignment accuracy within ±2 μm. The welding system recorded every force and pulse, ensuring complete traceability for each device.

The finished micro-welds underwent environmental testing; 1,000 cycles of heat cycling between −20°C and +60°C revealed no deterioration in mechanical performance or resistance. The joint resisted tensile pull stresses that were more than 30%

above specification, and biocompatibility testing verified that there were no leachables or cytotoxic reactions. The first-pass yield surpassed 98%, with almost 10,000 assemblies produced.

Resistance micro-welding's ability to consistently join ultrafine Pt–Ir wires without sacrificing the assembly's mechanical or electrical qualities or the implant's biocompatibility serves as justification for its use in this application. The procedure ensured a robust, small, and secure implantable pacemaker lead system by preventing contamination, permitting complete automation, and satisfying strict regulatory and medical safety standards.

6.5 CASE STUDY 4: OPTICAL FIBER COUPLING VIA FUSION SPLICING

6.5.1 Introduction to Optical Fiber Coupling and Fusion Splicing

In contemporary telecommunications, fusion splicing has emerged as the industry standard for optical fiber coupling. This procedure creates a continuous optical channel with little signal degradation by permanently uniting two optical fibers end-to-end through targeted melting, usually with the use of an electric arc. Extremely low signal loss and nearly nonexistent reflectance at splice sites are requirements for telecom networks. These demanding specifications are consistently met by fusion splicing, which makes it ideal for single-mode fiber applications where maintaining signal integrity over long distances is crucial for high-capacity data transfer.[24]

6.5.2 Technical Overview of Fusion Splicing

A controlled heat source, usually an electric arc, is used to melt two carefully prepared fiber ends together in fusion splicing, a precise thermal joining process. The fiber cores have to be perfectly aligned and energy applied such that the least amount of interruption, reflection, or scattering of light occurs when passing through the splice. Contemporary splicers put into use automated alignment methods with submicron precision, including core or profile alignment and Local Injection and Detection (LID). After the alignment, the glass is melted by an arc discharge for milliseconds, and capillary forces assist in the smooth fusion of molten ends to bring down the splicing losses to tolerances most commonly less than 0.02 dB.[25]

6.5.3 Process Steps in Fusion Splicing

Fiber ends are carefully prepared before actually beginning the fusion splicing process. Once the polymer coatings are removed so as to expose the naked silica core and cladding, the technician gently cleans the fiber, often by using lint-free wipes soaked with isopropyl alcohol, and then cuts the fiber end perpendicular to the axis using precise tools. These procedures minimize scattering errors and guarantee the best end shape for splicing. After the fiber ends are ready, they are inserted into the alignment mechanism of the fusion splicer. Depending on the machine, alignment is accomplished via LID, which injects and monitors light to fine-tune core centering, or optical core alignment, where imaging devices detect core placement from several directions. After alignment, a pre-fusion arc clears debris and warms the ends before the primary fusion arc melts and merges the fibers under controlled compression. The splice is mechanically delicate after the fusion and needs to be shielded. Before processing, protective sleeves—typically heat-shrink tubes reinforced with aramid—are put over the splice, and after fusing, they shrink into place. In high-reliability settings, such as underwater cabling, recoating is an optional procedure that uses UV-curable acrylate resin to restore the fiber's original diameter and mechanical robustness. Tensile integrity is then confirmed through proof testing.[26]

6.5.4 Signal Loss Considerations and Metro-Scale Requirement

Two important measures that determine the quality of a splice are insertion loss and back reflection. In high-performance single-mode fiber deployments, telecom networks aim for insertion losses as low as 0.02 dB, with joints aiming for far below 0.1 dB. In high-speed and long-distance networks, where dispersion and cumulative attenuation can greatly affect signal integrity, such low losses are essential. Attenuation rates for single-mode fibers operating at 1,300–1,550 nm may already be as low as 0.25–0.35 dB/km. In this situation, maintaining data speeds, reach, and dependability requires minimizing loss at splice sites while retaining signal purity.[25]

6.5.5 Equipment and Tooling

Essential subsystems that comprise fusion splicing equipment include high-precision cleavers, vision systems equipped with cameras or microscopes for alignment, automated splicer heads featuring movable V-grooves or actuators, arc generators, and heating units used for applying protective sleeves. Real-time loss estimators, ribbon capabilities, and integrated cleavers are features of advanced models. A large portion of the process—cleaving, alignment, fusion, testing, and protection—is automated by these technologies, which lowers operator error and boosts throughput. Ribbon splicers increase efficiency for dense fiber management applications by handling up to 12 fibers at once.[27]

6.5.6 Advantages and Challenges

Fusion splicing cannot match its low insertion loss, structural durability, and permanent optical continuity.[24] It circumvents the greater losses and misalignment hazards associated with mechanical splicing methods that use index matching or gels, which can create reflections of up to 0.3 dB. The equipment's complexity and expense, the need for hygienic surroundings, and operator training are its drawbacks. Quality training and standard operating procedures are necessary because errors in cleaving, alignment, or arc settings might produce splices that are not ideal, resulting in increased loss or even possible fiber damage.

6.5.7 Case Example: Long-Haul Telecom Backbone Deployment

A significant telecom provider started modernizing its nationwide backbone network in anticipation of the 400 Gbps long-distance transmission implementation. In order to ensure sub-0.02 dB splice loss targets at each joint, the upgrade required installing new single-mode fiber over thousands of kilometers and replacing aging mechanical splices with fusion splices.

Choosing premium ITU-T G.652.D single-mode fiber and corresponding pigtail modules was the first step in the project. With an emphasis on the most important processes—exact stripping, end-face cleaving, cleave quality inspection, automatic alignment selection, and parameter adjustment for the local environmental temperature and humidity—technicians received in-depth instruction in fusion splicing techniques.[28]

The splicing procedure began in the field by mechanically removing the buffer and coating from a 40 mm piece of fiber. After that, lint-free wipes and high-purity isopropyl alcohol were used to carefully clean the exposed fiber. Precision cleavers were used to cut each fiber end, resulting in end-face angles that were within 0.5°, as later verified by the splicer's inspection instruments. Operators noticed that the fiber surfaces were more susceptible to electrostatic dust during a prolonged dry season; thus, extra cleaning procedures were incorporated into the routine.[29]

The produced fibers were put into a fusion splicer, which offered the highest alignment precision thanks to its alignment system that used the profile alignment approach for single fibers. Any cleaves that did not meet the acceptable requirement were automatically detected and remade by the splicer, which also recorded real-time photos and checked the quality of the cleaves. The splicer used a pre-fusion arc to remove impurities and soften the ends after passive alignment. The primary fusion arc was then adjusted for local air pressure, which was a crucial adjustment because arc behavior changed in regions higher than 3,000 meters. For around two seconds, the fusion arc provided enough energy to completely melt the glass ends while minimizing core deformation or axial displacement. Operators documented the outcomes, while the splicers offered real-time splice loss estimation. The average insertion loss over 5,000 splices in the initial testing was 0.018 dB with a standard variation of 0.002 dB, showing consistency well within performance requirements. Splices that had a loss of more than 0.03 dB were marked for recleaving and re-splicing right away.

Heat-shrink sleeves reinforced with aramid were utilized to protect the splice sites after they were fused. After fusing, sleeves were pre-positioned and heated using built-in heaters to shrink them tightly. Recoating with UV-curable acrylate was also used to restore the fiber's original diameter and bend performance in high-reliability trunk portions intended for subsea or duct environments. A 1% elongation proof test verified that every recoated splice surpassed mechanical strength standards.[30]

To verify continuity and validate backscatter functionality, OTDR testing was performed on each splice. The optical integrity was confirmed by the OTDR trace, which revealed no loopbacks or reflection peaks at splice sites.

Test segments were deployed to replicate temperature extremes (–40°C to +70°C) with cycling for 500 cycles as part of the environmental testing. Under tensile stress and vibration, mechanical integrity held firm, while splice loss stayed constant. In order to create a comprehensive quality and traceability record that is necessary for certification to operator and regulatory standards, the fusion splicers recorded every parameter used in each splice.[31]

The average loss remained constant at 0.019 dB throughout the whole deployment, which included 10,000 fusion splices, and the first-pass yield surpassed 98.5%. Longer repeater spacing and cost reductions on pumping amplifiers and regeneration stations were made possible by reduced signal loss. By extending reach by 35%, the network improvement made it possible to install the new 400 Gbps channels over previously inaccessible areas.

Both performance and economics were used to support fusion splicing, especially the consistent achievement of <0.02 dB loss. Joints with high insertion loss would require more amplifiers or signal filtering, which would raise operating costs and power consumption. The physical joints' long-term dependability was also crucial because mechanical splices were not very durable, particularly in subterranean or submerged environments. When considering reduced signal degradation, extended maintenance intervals, and uptime reliability, fusion splicing has proven to be cost-effective despite its capital-intensive nature.[32]

This example showed how fusion splicing is an essential component of high-speed, high-capacity, and future-proof optical networks, not just a connecting technique.

6.5.8 Summary

The foundation of high-performance telecom network infrastructure is fusion splicing, which allows for high-bandwidth, long-distance optical transmission with little signal loss. With careful alignment, preparation, and regulated heat fusion, splice performance can reliably reach high standards (<0.02 dB loss). The aforementioned case study demonstrates both the technical difficulty and the strategic rationale—fusion splicing guarantees network scalability, longevity, and efficiency in contemporary fiber-optic systems.

6.6 CASE STUDY 5: BATTERY TAB WELDING BY ULTRASONIC MICRO-WELDING FOR JOINING THIN-FOIL CURRENT COLLECTORS IN LI-ION COIN AND POUCH CELLS

6.6.1 Introduction

Lithium-ion battery connectors' dependability and manufacturing feasibility are now more important than ever due to the shift to electrified vehicles and ubiquitous portable gadgets. The integrity of the current collector joints, which are where thin foils of copper or aluminum are attached to tabs, leads, or busbars, can determine whether a cell passes production testing, maintains low internal resistance over its lifetime, and withstands abuse conditions without catastrophic failure, even though cathode and anode chemistries frequently take center stage in discussions. Therefore, the field of battery tab welding intersects with industrial engineering, mechanics, and electrochemistry. Because it produces joints with very low electrical resistance and excellent fatigue behavior, maintains active material coatings and polymeric separators nearby, and produces solid-state bonds at low bulk temperatures, ultrasonic micro-welding has become the preferred technique for joining these delicate, highly conductive foils. This case study concludes with a lengthy, extremely detailed case example that traces a program from laboratory coin cells through pilot production of pouch cells. It explains the technical foundation and real-world application of ultrasonic micro-welding for joining thin aluminum and copper current collectors to tabs in coin and pouch cells.[33]

6.6.2 Ultrasonic Micro-Welding Fundamentals

Two or more metal layers are clamped together and energized by high-frequency vibratory motion that is transmitted through a sonotrode in the pressure-assisted, solid-state process known as ultrasonic micro-welding. Depending on the alloy and stack rigidity, the horn oscillates parallel to the faying interface, typically at 20–40 kHz, with tip amplitudes ranging from 10–40 micrometers. When paired with a standard clamping force, the applied oscillatory shear causes severe plastic deformation at asperity-scale contact zones, ruptures surface oxides, and disturbs adsorbed pollutants. While the bulk joint stays well below the melting point, localized frictional heating raises the temperature of these micro-contacts into a range where mechanical mixing and diffusion take place. The interface changes from discrete micro-junctions to a continuous metallic bridge network as asperities collapse and real contact area increases.[34] The procedure encourages dislocation production and dynamic recovery for face-centered cubic conductors, such as copper and aluminum, which further softens the near-surface layers and fosters intimate bonding without creating brittle intermetallic phases. Thin-foil mechanics and coating sensitivity limit the process window in battery tab welding. In energy-dense cells, anode and cathode foils are normally 6–20 μm thick for copper and 12–25 μm thick for aluminum. These foils contain active material coatings, such as high-voltage spinels or layered oxides on aluminum and graphite or silicon-dominant anodes on copper, which are held together by polymeric binders with thermal limits far lower than the metals' melting points. As long as the energy input and cycle time are strictly controlled, the active layers and the separator stack are protected because ultrasonic welding confines the majority of the heat to a shallow area within a few micrometers of the interface. Shrinkage porosity and thermal distortion are eliminated by the lack of bulk melting, and high-throughput manufacturing is made possible by the short cycle times, which are usually a few hundred milliseconds.[35]

6.6.3 Materials and Stack-Ups in Coin and Pouch Cells

In lithium-ion cells, copper serves as the negative current collector, while aluminum serves as the positive current collector. The "tabs" in coin cells used for laboratory testing are frequently either part of the lid design or are short nickel or aluminum coupons that are spot-joined to foil tails prior to crimping the cap. Tabs are elongated foils or laminated composites used in production pouch cells. They are typically made of copper or nickel-plated copper for the negative and aluminum for the positive. They are welded to multilayer stacks of collector foils and then routed out of the laminate pouch through a seal area that must withstand pull loads and heat sealing. Occasionally it is impossible to prevent dissimilar-metal junctions. For example, manufacturers may favor nickel-plated tabs because of their resistance to corrosion, solderability, or compatibility with busbars, while the collector is made of copper or aluminum. As long as the energies are adjusted to prevent the interface from becoming too old, ultrasonic welding can fuse these disparate combinations without producing troublesome intermetallic layers. Surface oxide is the main bonding barrier for aluminum stack-ups; native oxide is thinner for copper, but if horn textures are not controlled, work-hardening and smear might compromise uniformity. For insulation or strain relief, tab laminates may include polymer films or adhesives. These layers must be hidden or kept out of the weld zone since they absorb ultrasonic energy and can volatilize when heated by friction.[35]

6.6.4 Equipment Architecture and Process Control

An ultrasonic battery tab welder consists of a power source, a piezoelectric transducer stack, a booster, and a sonotrode or horn that contacts the tab side of the stack. The collector foils are supported by the opposing anvil. The stack is clamped by servo-electric or pneumatic actuators using standard force. In addition to power and duration, advanced welders also regulate closed-loop amplitude and real-time energy delivery. Many of them use displacement sensing to identify stack collapse and stop the cycle when a predetermined weld "signature" is reached. The asperity pitch and height of the micro-textures—knurls, pyramids, or linear grooves—machined into the horn faces for foil welding dictate how forcefully the horn grasps the top foil and how the shear is transferred. Too smooth a texture can encourage slippage without sufficient frictional heating, while too aggressive a texture can lacerate top foils. Energy coupling and tooling footprint are balanced by frequency selection: 20 kHz systems are typically used for thicker busbars and multilayer stacks, whereas 35–40 kHz systems are preferred close to sensitive separators because they offer finer energy control for fragile foils. Clamping force (tens to hundreds of newtons), amplitude (usually 12–30 μm peak-to-peak at 20–40 kHz), weld time (50–500 ms), and energy (1–20 J per weld for foil stacks, increasing with layer count) are some of the process parameters. Coin cells employ smaller targets but may execute several overlapping welds to spread current; for pouch tabs, the welded area is typically a rectangle that is one to several millimeters broad and several millimeters long.[36]

6.6.5 Metallurgical and Electrochemical Integrity

An ultrasonic weld in copper or aluminum has a different metallurgical nature than a fusion junction. Evidence of severe plastic deformation can be seen at the interface in the form of refined grains with an extended substructure oriented along the shear direction. Aluminum oxide films are broken up and dislocated; they can occasionally be seen as ribbons that have been extruded to the weld's edge. Although the lamellae at copper contacts are severely distorted, the metal lattice remains continuous throughout the bond. There is no cast structure and no resolidification flaws because the process never creates a liquid phase. The absence of intermetallic growth is crucial in dissimilar connections like copper-to-nickel or aluminum-to-nickel, where brittle phases would normally impair fatigue resistance. From an electrochemical perspective, the joint must prevent galvanic pairings in electrolyte-rich situations and impart very little series resistance.[37] To increase corrosion resistance without altering bulk conductivity, nickel-plated copper tabs are frequently used on the negative terminal. Barrier foils or aluminum tabs are recommended when aluminum and nickel must meet for positive tabs to prevent galvanic mismatch; if dissimilar contact cannot be avoided, the thin diffusion layer and tiny exposed area of ultrasonic welding minimize galvanic currents. Because metallic debris might migrate and result in soft shorts, residual particles produced by foil fretting must be reduced; process design includes housekeeping and efficient vacuum extraction close to the weld head.[38]

6.6.6 Mechanical Reliability, Electrical Performance, and Quality Assurance

Lap-shear and T-peel tests, which examine cohesive joint strength and delamination behavior, are used to assess mechanical performance. In service, tabs undergo temperature cycling and vibration; hence, high-cycle fatigue tests with modest displacement amplitudes are equally crucial. Either Kelvin clamp fixtures built into the welder or four-wire milliohm readings across defined probe spacings are used to evaluate electrical performance. With failure happening in the foil rather than at the interface, production targets for pouch cell tab joints typically call for peel strengths exceeding dozens of newtons per centimeter of weld width and extra resistance per junction of less than a few hundred micro-ohms. To identify out-of-family occurrences like inadequate clamping or oil contamination, nondestructive quality controls involve tracking weld power curves, energy delivery, dynamic resistance, and displacement profiles. Effective oxide disruption and the lack of delamination are confirmed by cross-section metallography. Since the formation of oxide films on aluminum can increase resistance over time if the initial bonding is insufficient, joints are also assessed for stability during accelerated aging at elevated temperatures and for heat rise under pulse loads in accordance with cell safety.[39]

6.6.7 Process Integration with Cell Assembly

Ultrasonic welding is frequently used in coin cells when fabricators incorporate custom lids with feedthroughs or when researchers need to attach pigtails to foils for external measurements. The cap's low mass and close closeness to the crimp area present a problem; tooling must provide stiff backing without warping the sealing surfaces. Ultrasonic welding is fully integrated in pouch cells: following the assembly of electrode stacks or jelly rolls, the foil tails are collected and positioned with the tab stock inside special fixtures that shield the separator edge from horn contact. Weld sequences are planned to prevent excessive local thinning and control heat accumulation. Process engineers coordinate seal temperatures and dwell times with weld sites and tab stack thickness because localized heat from sealing must not relax the ultrasonic welds when the tabs are routed through the pouch foil's seal area after tab welding. Lamination and heat sealing are essential to pouch manufacture, so any contamination by meltable polymers close to the weld zone must be carefully managed.[39]

6.6.8 Case Example: Development and Scale-Up of Ultrasonic Micro-Welds for Thin-Foil Current Collectors in Coin and Pouch Cells

6.6.8.1 Project Context and Objectives

Using coin-cell formulas from labs, a battery company set out to create a high-energy pouch cell platform. To optimize areal capacity, the original coin cells used 15 μm aluminum cathode collectors with high-loading coatings and 10 μm copper anode collectors. With tab groups soldered to low-resistance foils for module connections, the target pouch cell, rated at 30 Ah for automotive modules, would stack dozens of electrodes per polarity. Three strict requirements were established by the engineering team for the tab welds from the beginning: the joints must contribute less than 0.1 mΩ of series resistance per polarity, the peel failure mode must be "foil tear" instead of interfacial separation, and the process must be resilient to the anticipated variation in foil roughness, coating caliper, and residual solvent content. To maintain the mechanical qualities of the polyolefin separator, the team additionally enforced a thermal budget, which states that local temperatures at the separator edge next to the weld zone cannot rise above 80°C during any welding operation. The program would develop gradually. First, utilizing short tab coupons, coin-cell pigtail welding would act as a low-risk test bed for adjusting the ultrasonic settings for copper and aluminum foils. To assess routing through the pouch seal, a subscale pouch demonstrator would then incorporate tiny tab stacks. Lastly, a pilot-line production cell would confirm long-term stability, tool wear, and throughput.[40]

6.6.8.2 Laboratory Coin-Cell Phase: Establishing the Weld Window

The current collectors were represented by copper and aluminum foil strips at the start of the laboratory phase. These strips were trimmed to 40 mm in length and 10 mm in width, leaving the uncoated ends for welding. A 50 μm copper tab and 10 μm copper foil were used for the negative side, and a 100 μm aluminum tab and 15 μm aluminum foil were used for the positive side. To provide more precise amplitude control on these fragile stacks, a tabletop ultrasonic welder with detachable heads operating at 35 kHz and 40 kHz was used. To overcome the more stubborn oxide, the horn faces were produced with micro-pyramidal patterns of 100 μm pitch and 25 μm height for copper and a somewhat sharper 80 μm pitch texture for aluminum. The group created a factorial experiment with clamping force ranging from 60 to 160 N, weld duration ranging from 80 to 260 ms, and amplitude changing between 12 and 22 μm. To prevent work-hardening artifacts from previous runs, three copies of each condition were welded on new foil.[41]

T-peel strips were ready for mechanical testing as soon as the welding was completed, and contact resistance was measured using a four-wire micro-ohm meter. For microscopy, cross sections of a few chosen joints were polished to 0.05 μm and placed in cold epoxy. Peak temperatures were recorded during welding using thermocouples positioned 1 mm from the weld zone to guarantee that the separator thermal budget could be adhered to in the pouch configuration of the future. Characteristic interactions were found in the data. At low amplitude, resistance stayed over 0.4 mΩ for copper and 0.8 mΩ for aluminum, and even long weld durations were insufficient to collapse asperities. When oxide rupture became widespread and the real contact area grew, increasing the amplitude to 18–20 μm significantly decreased resistance. However, local thinning and early tearing in the copper foils were produced by excessive force at high amplitude, as shown by shear bands that could be seen under a microscope. For copper, the sweet spot was 18 μm amplitude, 110–130 N force, and 140–180 ms time. This resulted in peel forces that split the foil instead of the interface and mean resistance < 0.08 mΩ. To maintain low-heat input, aluminum needed a slightly larger amplitude, about 20–22 μm, at comparable forces but shorter periods; resistance dropped below 0.1 mΩ, and foil peel failure was constant. With a 1 mm offset, thermocouples recorded peak temperatures below 55°C, well below the anticipated separator limit. The results were confirmed by metallography. Whereas aluminum sections displayed fragmented oxide dispersoids extruded toward the weld perimeter and a refined subgrain structure consisting of 0.5–1.5 μm features, copper joints displayed a well-consolidated interface with smeared lamellae and evidence of dynamic recrystallization close to the faying surfaces. There was no evidence of voiding or delamination. The thickness was maintained because the horn textures did not pierce the collector foil and instead created shallow indentations on the tab side. With resistance standard deviations less than 0.01 mΩ, the variability stayed low after ten sets of strips were subjected to the optimal settings on various days and with various foil lots. This robustness suggested a viable weld window for transferring to pouch geometries.[41]

6.6.8.3 Subscale Pouch Demonstrator: Managing Stack-Ups and Heat Paths

The complexity of multilayer stacks and proximity to separators were introduced in the following step. A tiny stack of five negative and five positive electrodes per side, each with 10 mm foil tails reaching to the tab region, was housed in a subscale pouch fixture. The positive tab was made of 100 μm aluminum, while the negative tab laminate was made of 100 μm copper foil with a 10 μm nickel flash layer close to the free end to match a nickel-plated busbar later in the module assembly. The goal was to use an anvil with a recess to maintain the separator edge 1.5 mm from the horn contact zone and to fuse the five copper tails to the copper tab in a single ultrasonic cycle for aluminum. The team shifted to a 20 kHz power head to produce greater amplitudes and more aggressive oxide disruption, especially on the aluminum side, because five layers enhance stack rigidity. The single-foil recipes were used as a starting point for the trials, and their amplitude was gradually increased to account for the stack's decreased compliance. In order to guarantee consistent contact between all layers, the clamping force was raised to 300–350 N. A distinctive collapse curve, where the stack compressed by 25–40 μm in the first 80–120 ms and subsequently stabilized, was connected with good metallurgical bonding across all layers, making real-time displacement monitoring a crucial signal. The bottom foils were not engaging if displacement plateaued too early; the top foils ran the risk of over-thinning if displacement increased further into the cycle. A 20 kHz, 24 μm amplitude, 320 N force, and 180 ms duration resulted in homogenous joints on the copper side. Resistance evaluated using busbar-style Kelvin clamps averaged 0.05 mΩ per joint, and cross sections cut at three lateral points demonstrated consistent bonding across the five sheets. Amplitude and texture had to be carefully balanced while working with aluminum. Under etched microscopy, isolated unbonded islands were apparent due to sporadic slides in the bottom foil in early runs with the copper horn texture. By switching to a sharper horn micro-texture designed specifically for aluminum, the problem was resolved and the duration was reduced to 140 ms at 22 μm amplitude and 330 N force, resulting in resistance that was close to 0.07–0.09 mΩ. Peak surface temperatures of about 70°C for aluminum and 60°C for copper were observed by infrared micro-pyrometry directed at the tab edge; implanted micro-thermistors at the separator edge consistently maintained temperatures below 65°C, meeting the thermal guardrail. Coupons taken from the subscale tab region were subjected to mechanical peel testing, which produced force-displacement curves with a steady peak and sawtooth tearing as foil layers failed one after the other. As the number of layers increased, the energy under the curve—a stand-in for labor to separate—rose linearly, suggesting that each layer made an equal contribution to joint toughness. The absence of brittle intermetallics and the plastically conformed interface shape are likely responsible for the fatigue robustness, as evidenced by high-frequency fatigue testing at 20 Hz and ±0.25 mm tab deflection that lasted for one million cycles without resistance drifting over 5 μΩ.

6.6.8.4 Pilot-Line Pouch Cells: Tooling Durability, Throughput, and Statistical Control

After achieving subscale success, the group put in place a pilot line to manufacture 30 Ah pouch cells with full-length tabs that are appropriate for module assembly and 24 layers per polarity. In order to weld negative and positive tabs simultaneously, the ultrasonic station was set up with two heads. Each head was fed by an automatic stack loader that aligned foil tails with an accuracy of ±0.15 mm. A medium knurl for copper and a sharper knurl for aluminum were the two face textures required; titanium horns were made to minimize wear and reduce mass. Anvils included spring-loaded micro-support pins beneath the outermost foils to guarantee equal contact across the width and vacuum apertures to remove particles during welding. Process engineers used energy-based termination and closed-loop amplitude control to standardize weld quality across foil hardening variances caused by rolling lot variations. A multidimensional signature was produced by logging the amplitude, power, energy, collapse displacement, and dynamic resistance for each weld cycle based on voltage and current. Each signature was compared to a principal-component-based model trained on hundreds of good welds using a real-time statistical process control (SPC) engine. Depending on the extent of the deviation, each weld that fell outside the control ellipse triggered either an automatic reweld or a stoppage for inspection. When matched with upstream stacking and downstream sealing activities, the station's 0.8 seconds for each weld, including load/unload, allowed for more than 2,000 cells per shift per head. A gradual increase in the energy needed to accomplish the desired collapse was the first indication of tool wear on the aluminum horn texture, where oxide-rich smearing polished the asperities and decreased grip. Horn resurfacing was initiated by predictive maintenance rules when the SPC engine detected a consistent change in signatures or after 25,000 welds. Although copper horns were turned proactively to avoid catastrophic wear, they had a longer lifespan—more than 40,000 welds. Every 500 welds, anvils were cleaned to get rid of compacted foil fines that can change contact. In order to identify tab contributions, the results of end-of-line electrical tests were analyzed by comparing tabs welded at various widths that were regularly planned in the build plan. Internal resistance was measured at 1 kHz and using DC pulse methods. With a narrow spread, the tab-related resistance contribution stayed below 0.18 mΩ per polarity. Thermal imaging revealed hot patches that were consistent with busbar geometry rather than the weld itself, and abuse tests that pulsed 10C current for five seconds resulted in temperature increases at the tab of less than 8°C. Metallurgical stability was confirmed by the fact that cells aged for eight weeks at 60°C did not exhibit any discernible increase in tab resistance.[42]

6.6.8.5 Failure Analysis, Edge Cases, and Design Feedback

To strengthen the process, the team looked into three groups of anomalies notwithstanding the approach's general effectiveness. First, lubricant contamination from an upstream foil slitting process was identified as the cause of intermittent high-resistance outliers. Although the ultrasonic technique is sensitive to viscous coatings that prevent frictional heating, it is tolerant of thin oxides.[43] Two mitigating strategies were used: a poka-yoke interlock stopped welding when dynamic resistance at cycle start was abnormally low, which is a symptom of lubricated slip, and a plasma cleaning phase was included for tab areas. Second, sporadic "horn kiss" marks close to the separator edge indicated infrequent stack loader mis-picks that let a foil stick out; to increase foil tail recognition, the loader's vision system was retrained with a darker backdrop and edge-enhanced lighting. Third, bend testing revealed an intermittent peeling mode at the outer border of the aluminum tab, indicating local underbonding when horn pressure deteriorated close to the corners. Without altering global settings, the problem was resolved with a redesigned horn that had a modest crown and increased edge pressure by a few percent. This equalized energy deposition. The cell architects received input on their designs. They approved a nickel flash on the outer 5 mm of the copper tab to ensure corrosion resistance without affecting the ultrasonic interface region, specified a tighter roughness band on aluminum foil to reduce variability in oxide disruption energy, and widened the tab neck by 0.5 mm to reduce current density at the transition. To prevent any polymer from intruding into the ultrasonic zone—even minute amounts have been shown to dampen energy and produce interfacial "cold spots"—the pouch laminate provider modified the adhesive bleed control.[44]

6.6.8.6 Electrochemical Correlation and System-Level Justification

Pilot-line cells' electrochemical cycling validated the significance of the mechanical and electrical measurements. For moderate rates, cells with ultrasonic tabs retained the same amount of capacity as baseline cells. However, because of their lower series resistance, ultrasonic-tab cells provided 10–15 mV more voltage stability for high-power pulses. Near the tab region, thermal gradients during strong discharge were diminished, resulting in a slower rate of impedance development with aging. Tests conducted at the module level showed that the constant tab resistance decreased pack balance variation from cell to cell, saving top-of-charge equalization procedures several minutes. By allowing the single-cycle joining of multilayer stacks without the need for filler metals, shielding gases, or lengthy dwell times, ultrasonic micro-welding economically justified its capital and engineering expenditures.

Ultrasonic welding decreased rework by removing splash and electrode sticking on thin foils, and it increased tooling life by orders of magnitude because the sonotrode did not deteriorate via arcing as it was being worn, unlike resistance spot welding used in a previous generation of cells. Additionally, the procedure demonstrated inherent scalability: the addition of parallel heads resulted in a linear increase in throughput, and the closed-loop signatures enabled lights-out production by means of automated decision-making.

6.6.8.7 Final Validation

The development arc produced several timeless lessons. First, horn face engineering holds equal importance to parameter tuning; it is crucial to match the alloy, layer count, and surface condition with the micro-texture. Second, in thin-foil stacks, displacement and dynamic resistance signatures are more sensitive than time-or-energy-only measures to indicate the health of the weld. Third, coordinating upstream and downstream processes prevents latent problems: precise stack handling, clean slitting, and adhesive control ensure the weld window remains intact. Lastly, scale-dependent stiffness and heat routes require parameter adjustments when moving recipes from coin-scale studies to pouch production; nonetheless, the basic physics is retained, enabling a methodical extrapolation. Over the course of two weeks, the final validation run generated 5,000 cells with process capability indices for resistance and peel strength comfortably over 1.67 and without any weld-related field failures. Following creation, random cell tear-downs verified pristine interfaces and the absence of separator thermal distress. The SPC signature library was required as a valuable resource for upcoming plants, and the manufacturer sealed the ultrasonic method as the platform's standard.

6.6.9 Conclusion

In lithium-ion coin and pouch cells, ultrasonic micro-welding has been demonstrated to be the most efficient method for welding battery tabs. This is especially true when the junction in question involves thin aluminum and copper current collector foils. Its solid-state method preserves the fragile microstructures and polymeric components surrounding the joint by combining controlled frictional heating with intense, localized plastic deformation to establish intimate metallic connections without melting. By carefully controlling amplitude, force, time, horn texture, and fixture design, the technique produces remarkably low electrical resistance and robust mechanical performance. It is also easily scalable from laboratory coin cells to high-throughput pouch production. The extended case study shows how performance and manufacturability may be achieved with a carefully designed weld window that is verified by metallography, electrical measurements, thermal protections, and SPC. Ultrasonic micro-welding offers the physics, accuracy, and practicality needed to link the electrochemistry of the cell to the realities of the power system in a world where every milliohm counts and every gram of mass is carefully examined.

6.7 CASE STUDY 6: MICROFLUIDIC CHIP SEALING USING DIFFUSION BONDING FOR HERMETIC GLASS/POLYMER MICROCHANNELS IN LAB-ON-A-CHIP DEVICES

6.7.1 Introduction

Microfluidics had an impact on analytical chemistry, biological diagnostics, and environmental monitoring, thanks to the unique ability of these systems to manipulate various fluids through channels usually less than one millimeter in diameter. A lab-on-a-chip system integrates various processes: preparation of the sample, reaction, separation, and detection on a far smaller platform than those known from traditional laboratories. Microfluidic systems demand adequately well-defined channel geometry and sealing to ensure their fluidic control is precise and devoid of leakage, contamination, or evaporation.[45]

The sealing of nanochannels or microchannels etched onto the substrate is an important stage in manufacturing microfluidic devices. The sealing technique would not alter the surface characteristics or structural integrity of the channels, guarantee chemical compatibilities with operating fluids, and withstand mechanical or thermal stresses that might evolve through their working state. Diffusion bonding stands as one of the most reliable joining techniques to produce hermetic seals that are mechanically strong and free from contamination. Other techniques can be adhesive bonding, thermal fusion, plasma-aided sealing, and ultrasonic welding. Diffusion bonding is the solid-state joining process that facilitates the atomic diffusion across an interface by pressing two mating surfaces together at a temperature high enough to allow diffusion but below the melting point. Therefore, the joint is monolithic and has bulk properties of the joined material. In the microfluidic field, diffusion bonding is very attractive to seal glass-glass, glass-polymer, and some polymer-polymer combinations since it avoids introducing adhesives or other foreign materials into the microchannels. This maintains the optical clarity and chemical purity essential for analytical purposes. Considered is this case study: the diffusion bonding for the hermetic sealing of glass/polymer microchannels found in microfluidic devices for biological diagnostics. It reviews the scientific principles and material considerations behind the process and looks into industrial applications and performance evaluation. It then presents an in-depth practical example of diffusion bonding investigations into a lab-on-a-chip for quick pathogen detection, describing the parameters and difficulties faced during the process as well as the results.[46]

6.7.2 Fundamentals of Diffusion Bonding

Diffusion bonding is a solid-state joining method that can hold two mating surfaces for a period at high temperature and pressure to allow atomic diffusion to occur across the interface. In contradistinction to fusion welding, this technique uses atomic mobility to remove interface barriers rather than melting the underlying materials. The mechanism of bonding involves the following stages. Due to surface roughness, contact first occurs between discrete asperities when the two surfaces are forced together. The real contact area increases as plastic deformation, localized to these contact zones, is generated by the applied pressure and temperature. Atomistic transport across the interface can thus evolve through atomic diffusion mechanisms such as lattice diffusion and grain boundary diffusion, resulting in metallurgical or molecular continuity. Pore elimination and shrinkage occur in the last step, creating a bond that is nearly as strong as the bulk material. The process parameters in glass/polymer bonding for microfluidics need to be carefully adjusted. Many polymers have substantially lower melting temperatures than glass, which has a rather high thermal expansion coefficient. For bonding, the temperature has to be low enough for the polymer to avoid thermal degradation, distortion, or softening; it has to be high enough to undergo diffusion and surface wetting, and surface preparation in the form of chemical activation or plasma cleaning must be implemented so surfaces are high in energy and free from contaminants.

6.7.3 Challenges in Sealing Glass/Polymer Microchannels

Glass and polymers are examples of different materials that pose several technical difficulties upon bonding. First, bond delamination or microchannel distortion may occur due to residual strains resulting from the mismatch of the coefficient of thermal expansion after cooling. Second, consideration should be made for thermal management, as polymers may not withstand the high temperature that is usually required for diffusion bonding. Third, very clean and flat surfaces are required to optimize intimate contact and diffusion to produce hermetic sealing without the use of adhesives. Under pressure, the microchannels themselves can be distorted: excessive bonding force could cause the channels either to collapse or to change cross-sectional profile, which may adversely affect the flow rates of fluids, their mixing efficacy, and analytical precision. For lab-on-chips, optical clarity is crucial, particularly for optical detection methods such as absorbance, fluorescence, or Raman spectroscopy. Any blemishes, bubbles, or microvoids at the contact would reduce the accuracy of the measurement by light scattering. Advanced diffusion bonding uses surface activation technologies, such as oxygen plasma treatment, to enhance adhesion at lower pressures and temperatures by increasing surface energy. Precision spacers and alignment jigs are often used during bonding to prevent microchannel collapse.[47]

6.7.4 Material Science Considerations

The choice of materials profoundly influences the feasibility of diffusion bonding.

6.7.4.1 Glass Substrate

For microfluidics, glass has its place as a material because it's chemically resistant, mechanically rigid, and optically transparent. The soda-lime and borosilicate varieties tend to be common because of their relatively cheaper price and wet etching process compatibility. Glass is thus a stable backbone that resists swelling and warping under wet bond conditions.

6.7.4.2 Polymer Cover Layer

Polymers like cyclic olefin copolymer (COC), cyclic olefin polymer (COP), poly(methyl methacrylate) (PMMA), and polycarbonate (PC) are being utilized. COC and COP have an edge when it comes to biomedical uses because they have high optical transmission abilities and low water absorption properties, thereby making them biocompatible. This trait also makes them amenable to scalable production, unlike glass, which in turn has a higher processing cost.

6.7.4.3 Challenges with Hybrid Bonding

Glass and polymer also have very different thermal expansion coefficients (CTE). The CTE of glass is ~3–9 × 10^{-6}/°C, whereas that of polymers such as COC is ~60–70 × 10^{-6}/°C. Differential thermal expansion upon bonding can lead to residual stresses that degrade the bond or cause distortion in microchannel geometry. Careful temperature choice and controlled cooling rate are thus essential.

6.7.5 Diffusion Bonding Process Workflow

The bonding process involves several carefully controlled steps:

6.7.5.1 Surface Preparation

Both glass and polymer surfaces have to be thoroughly cleaned of dust, organic residues, and contaminants. Glass is generally cleaned with the RCA process (ammonium hydroxide, hydrogen peroxide, and water) followed by rinsing with DI water. The polymer is cleaned with isopropanol or acetone and dried with nitrogen.

6.7.5.2 Surface Activation

Oxygen plasma treatment is applied to both surfaces. Plasma introduces polar functional groups such as hydroxyl (–OH) and carbonyl (–C=O), increasing surface energy and hydrophilicity. This not only improves wettability but also provides sites for hydrogen bonding and possible covalent linkage formation at the interface.

6.7.5.3 Alignment

The glass substrate with etched channels is aligned with the polymer cover sheet under a stereomicroscope. Precision alignment is critical, especially when integrating inlet/outlet ports or multilayer channel networks.

6.7.6 Bonding Parameters

Bonding occurs in a precision press with flat platens, ensuring uniform load distribution. For glass/polymer systems:

- Temperature: typically 100°C–130°C (slightly below polymer Tg).
- Pressure: 0.2–1 MPa, depending on channel geometry.
- Time: 30–90 minutes.

6.7.6.1 Cooling

Controlled cooling under constant pressure minimizes residual stresses from thermal mismatch. Rapid cooling may induce warpage or interfacial cracks.

6.7.6.2 Interfacial Chemistry and Bond Formation

The bonding mechanism involves several interacting forces:

- Van der Waals Forces: These provide initial attraction at very short distances (~0.3–0.5 nm).
- Hydrogen Bonding: Plasma-activated glass presents hydroxyl groups that can hydrogen bond with oxygen-containing groups on the polymer.
- Covalent Bonding: In some cases, condensation reactions between hydroxyl groups on glass and polymer produce Si–O–C linkages, enhancing stability.
- Polymer Chain Interpenetration: At bonding temperatures near Tg, polymer chains gain mobility and partially interpenetrate into nanoscale roughness on the glass surface.

6.7.7 Case Example: Diffusion Bonding for Hermetic Sealing of Glass-Polymer Microchannels in a Lab-on-a-Chip for Biomedical Diagnostics

6.7.7.1 Device Design and Functional Requirements

In this case study, the microfluidic device was made for quick nucleic acid amplification testing (NAAT), which finds harmful DNA/RNA sequences in blood plasma samples. With a footprint of an area that was just 25 × 25 mm^2, it essentially combined microchannels, reaction chambers, and detecting zones to serve as a portable diagnostic tool that suited POC environments. There were extremely high functional requirements with this device. First, because the assay used trace-level nucleic acids that were quickly harmed by exposure to the environment, the microchannels needed to be hermetically sealed to stop sample leakage or contamination. Second, the apparatus had optical transparency for fluorescence-based nucleic acid amplification detection. Thirdly, sealed microchannels had to withstand a maximum operating pressure of three bars during reagent mixing and pumping due to the step requirements of the assay. Finally, the substrate material ensuring device workings should not hinder enzymatic activities like PCR and must be chemically inert and biocompatible. Hence, bonding by adhesives had been rejected due to possible contamination of reagents and autofluorescence, producing a background signal. The risk of channel collapse made the thermal fusion of polymer materials an inadequate bonding technique. Diffusion bonding between a COC cover plate and a patterned glass substrate was therefore chosen as the sealing technique.[48]

6.7.7.2 Substrate and Channel Fabrication

Borosilicate glass was chosen as the fundamental substrate due to its chemical resistance, low autofluorescence, and thermal stability up to 500°C. Microchannels were made using photolithography and wet etching in hydrofluoric acid (HF), and microchannels with a depth of 50 μm and a width ranging from 80 μm (narrow capillaries) to 400 μm (mixing chambers) were fabricated. Atomic force microscopy (AFM) confirmed that the glass channels' surface roughness was less than 10 nm RMS, which is essential for creating close surface contact during bonding. The COC (Topas® 6013) cover plate was chosen due to its high glass transition temperature (Tg = 134°C), low water absorption (<0.01%), and optical clarity in the visible spectrum. To create inlet/outlet ports that were in line with the underlying glass channels, the 300 μm thick COC sheet was laser-microdrilled.

6.7.7.3 Surface Activation Before Bonding

The mismatch between surface chemistry and thermal characteristics is one of the main obstacles to attaching glass to COC. Whereas COC is a hydrophobic, nonpolar polymer with a low surface energy, glass has a hydrophilic surface that is abundant in silanol (Si–OH) groups. Direct bonding causes interfacial gaps and weak van der Waals adhesion if left untreated. The surfaces were treated with oxygen plasma to get around this.[48]

6.7.7.4 Challenges Encountered and Mitigation Strategies

In spite of this success, there was difficulty along the way. One of these difficulties was channel deformation underbonding, which resulted when the melted COC penetrated the glass-etched channels under high pressure or temperature, reducing flow cross sections. By inserting a sacrificial filler (a solution of polyethylene glycol) into the channels during bonding, this issue was temporarily mitigated. The sacrificial filler effectively resisted deformation and was eventually dissolved after the bonding process. Another challenge was the residual stress caused by thermal mismatch. Differential contraction resulted in crack formation at the contact if cooling rates exceeded 10°C/min. Success, therefore, rested on careful thermal ramping. Finally, 500 PCR thermocycles (60°C to 95°C cycling) were conducted on the chips to ensure their long-term stability under cyclic thermal loading. While adhesive-bonded devices exhibited delamination after approximately 200 cycles, diffusion-bonded chips remained intact throughout.

6.7.8 Summary

This case study illustrates the successful application of diffusion bonding for the hermetic sealing of glass/polymer microchannels in a lab-on-a-chip for rapid pathogen identification. To overcome the challenges of interface chemistry and thermal mismatch, the process needed meticulous material optimization, surface activation, and thermal–mechanical bonding conditions. Practical diagnostic applications were able to exploit the resulting devices because of their excellent hermeticity, mechanical strength, optical performance, and biochemical compatibility. Diffusion bonding is the gold standard of sealing in hybrid microfluidic devices due to it being superior to adhesive or thermal fusion methods, particularly when pressure-resistant devices are needed for biological assays, optical transparency, and contamination-free environments.

REFERENCES

[1] Abi Raad, E., Vorländer, M. 2023. Acoustic monitoring of weld strength in ultrasonic metal welding by tracking welding stages. *Journal of Manufacturing Processes* 101, pp. 1055–1064.

[2] Amanat, N., James, N.L., McKenzie, D.R. 2010. Welding methods for joining thermoplastic polymers for the hermetic enclosure of medical devices. *Medical Engineering & Physics* 32(7), pp. 690–699.

[3] An, B., Zhou, H., Cao, J., Ming, P., Persic, J., Yao, J., Chang, A. 2023. A review of silver wire bonding techniques. *Micromachines* 14(11), p. 2129.

[4] Arweiler-Böllert, S., Liesegang, M., Beck, T., Jungbluth, J., Schmeer, S. 2023. Relation between interface geometry and tensile shear strength of ultrasonically welded joints. *Journal of Materials Engineering and Performance* 32(22), pp. 10469–10485.

[5] Balz, I., Raad, E.A., Rosenthal, E., Lohoff, R., Schiebahn, A., Reisgen, U., Vorländer, M. 2020. Process monitoring of ultrasonic metal welding of battery tabs using external sensor data. *Journal of Advanced Joining Processes* 1, p. 100005.

[6] Bisping, J., Jeschke, P. 2022. Detecting multidimensional pareto fronts during the optimization of a return channel within a centrifugal compressor. *Journal of Turbomachinery* 144(4), p. 041003.

[7] Cao, J., Zhang, J., Wu, B., Tang, H., Lv, C., Song, K., Yang, G., Cui, C., Gao, Y. 2021. Study on manufacturing technology of Ag-8.5Au-3.5Pd fine alloy wire. *Micromachines* 12(8), p. 938.

[8] Chen, Q., Li, S., Liu, Y., Zhang, X., Tang, Y., Chai, H., Huang, Y. 2020. Size-controllable Fe-N/C single-atom nanozyme with exceptional oxidase-like activity for sensitive detection of alkaline phosphatase. *Sensors and Actuators B: Chemical* 305, p. 127511.

[9] Czerny, B., Khatibi, G. 2020. Interface characterization of Cu Cu ball bonds by a fast shear fatigue method. *Microelectronics Reliability* 114, p. 113831.

[10] Eto, M., Araki, N., Yamada, T., Klengel, R., Klengel, S., Petzold, M., Sugiyama, M., Fujimoto, S. 2020. Effects of alloying elements in high reliability copper wire bond material for high temperature applications. *Microelectronics Reliability* 114, p. 113819.

[11] Eto, M., Araki, N., Yamada, T., Sugiyama, M., Fujimoto, S. 2021. Influence of post-bonding heating process on the long-term reliability of Cu/Al contact. *Microelectronics Reliability* 118, p. 114058.

[12] Eto, M., Araki, N., Yamada, T., Sugiyama, M., Fujimoto, S. 2021. Microstructural characterization of alloyed palladium coated copper wire under high temperature. *Microelectronics Reliability* 120, p. 114125.

[13] Gester, A., Sprigode, T., Wagner, G. 2025. Novel approach for in-line process monitoring during ultrasonic metal welding of dissimilar wire/terminal joints based on the thermoelectric effect. *Welding in the World* 69(2), pp. 363–372.

[14] Gester, A., Wagner, G., Pöthig, P., Bergmann, J.P., Fritzsche, M. 2022. Analysis of the oscillation behavior during ultrasonic welding of EN AW-1070 wire strands and EN CW004A terminals. *Welding in the World* 66(3), pp. 567–576.

[15] He, L., Zhang, M., Shams-Ansari, A., Zhu, R., Wang, C., Marko, L. 2019. Low-loss fiber-to-chip interface for lithium niobate photonic integrated circuits. *Optics Letters* 44(9), p. 2314.

[16] Hong, L., Zhang, C., Wahlen, J., Antonio-Lopez, J.E., Amezcua-Correa, R., Markos, C., Wang, Y. 2024. High energy and narrow linewidth N2-filled hollow-core fiber laser at 1.4 μm. *Journal of Lightwave Technology* 42(16), pp. 5645–5649.

[17] Kim, S., Park, J., So, S., Ahn, S., Choi, J., Koo, C., Joung, Y.-H. 2019. Characteristics of an implantable blood pressure sensor packaged by ultrafast laser microwelding. *Sensors* 19(8), p. 1801.

[18] Li, J., Zillner, J., Balle, F. 2023. In-depth evaluation of ultrasonically welded Al/Cu joint: Plastic deformation, microstructural evolution, and correlation with mechanical properties. *Materials* 16(8), p. 3033.

[19] Lin, C.-T., Lin, C. 2020. Multi-response optimization of wire bonding process for evaluating alternative wire material. *Microelectronics Journal* 106, p. 104925.

[20] Luo, H., Gan, D., Gama, M., Tu, J., Yao, F., Zhang, Q., Ao, H., Yang, Z., Li, J., Wan, Y. 2020. Interpenetrated nano- and submicro-fibrous biomimetic scaffolds towards enhanced mechanical and biological performances. *Materials Science and Engineering: C* 108, p. 110416.

[21] Ma, Y., Li, J., Dong, F., Yu, J. 2021. Power cycling failure analysis of double side cooled IGBT modules for automotive applications. *Microelectronics Reliability* 124, p. 114282.

[22] Majumdar, A., Mukherjee, T., Gangopadhyay, S. 2025. Estimation of Petermann I, Petermann II spot sizes for Kerr-type nonlinear dispersion-shifted along with dispersion-flattened optical fibers in fundamental mode guidance applying a simple but accurate formalism. *Journal of Nonlinear Optical Physics & Materials* 34(01).

[23] Müller, F.W., Chen, C.-Y., Schiebahn, A., Reisgen, U. 2023. Application of electrical power measurements for process monitoring in ultrasonic metal welding. *Welding in the World* 67(2), pp. 395–415.

[24] Müller, F.W., Schiebahn, A., Reisgen, U. 2022. Quality prediction of disturbed ultrasonic metal welds. *Journal of Advanced Joining Processes* 5, p. 100086.

[25] Murzinova, M.A., Shayakhmetova, E.R., Mukhametgalina, A.A., Sarkeeva, A.A., Nazarov, A.A. 2023. Local plastic deformation and quality of Cu-Cu joints obtained by ultrasonic welding. *Metals* 13(10), p. 1661.

[26] Ni, Z.L., Liu, Y., Ye, F.X. 2022. Formation of ultrasonic spot welded Cu/Cu NPs/Cu joints and the mechanical properties and electrical resistance. *Materials Science and Engineering: A* 845, p. 143251.

[27] Ni, Z.L., Yang, J.J., Ye, F.X. 2020. Microstructure and mechanical properties of copper to nickel ultrasonic spot welds. *Materials Science and Engineering: A* 796, p. 140207.

[28] Park, J.-H., Song, Z., Lee, G.-Y., Jeong, S.-M., Kang, M.-J., Pyun, J.-C. 2019. Hypersensitive electrochemical immunoassays based on highly N-doped silicon carbide (SiC) electrode. *Analytica Chimica Acta* 1073, pp. 30–38.

[29] Rahim, K., Mian, A. 2017. A review on laser processing in electronic and MEMS packaging. *Journal of Electronic Packaging* 139(3), p. 030801

[30] Rakshit, R., Majumdar, A., Maiti, S., Gangopadhyay, S. 2022. Influence of Kerr nonlinearity on single-mode dispersion-shifted and dispersion-flattened directional couplers: analysis by a simple but accurate method. *Optical and Quantum Electronics* 54(2), p. 118.

[31] Roshanghias, A., Rodrigues, A., Schwarz, S., Steiger-Thirsfeld, A. 2020. Thermosonic direct Cu pillar bonding for 3D die stacking. *SN Applied Sciences* 2(6), p. 1091.

[32] Shi, B., Marra, G., Feng, Z., Sakr, H., Hayes, J.R., Fokoua, E.R.N., Ding, M., Poletti, F., Richardson, D.J., Slavik, R. 2022. Temperature insensitive delay-line fiber interferometer operating at room temperature. *Journal of Lightwave Technology* 40(16), pp. 5716–5721.

[33] Shi, B., Zhang, C., Kelly, T., Wei, X., Ding, M., Huang, M., Fu, S., Poletti, F., Slavík, R. 2024. Splicing hollow-core fiber with standard glass-core fiber with ultralow back-reflection and low coupling loss. *ACS Photonics* 11(8), pp. 3288–3295.

[34] Shin, S., Nam, S., Yu, J., Park, J., Kim, D. 2021. Ultrasonic metal welding of multilayered copper foils to nickel-plated copper sheet in lithium-ion battery cell. *Metals* 11(8), p. 1195.
[35] Singh, G., Mamat, O. 2011. Effect of conduction pre-heating in Au-Al thermosonic wire bonding. *Journal of Surface Engineered Materials and Advanced Technology* 01(03), pp. 121–124.
[36] Suslov, D., Komanec, M., Numkam Fokoua, E.R., Dousek, D., Zhong, A., Zvánovec, S., Bradley, T.D., Poletti, F., Richardson, D.J., Slavík, R. 2021. Low loss and high performance interconnection between standard single-mode fiber and antiresonant hollow-core fiber. *Scientific Reports* 11(1), p. 8799.
[37] Taylor, C.S., Chen, R., D'Sa, R., Hunt, J.A., Curran, J.M., Haycock, J.W. 2021. Cost effective optimised synthetic surface modification strategies for enhanced control of neuronal cell differentiation and supporting neuronal and Schwann cell viability. *Journal of Biomedical Materials Research Part B: Applied Biomaterials* 109(11), pp. 1713–1723.
[38] Volpe, A., Di Niso, F., Gaudiuso, C., De Rosa, A., Vázquez, R.M., Ancona, A., Lugarà, P.M., Osellame, R. 2015. Welding of PMMA by a femtosecond fiber laser. *Optics Express* 23(4), p. 4114.
[39] Wang, C., Yu, R., Debord, B., Gérôme, F., Benabid, F., Chiang, K.S., Xiao, L. 2021. Ultralow-loss fusion splicing between negative curvature hollow-core fibers and conventional SMFs with a reverse-tapering method. *Optics Express* 29(14), p. 22470.
[40] Wang, T., Li, Y., Zhang, Y., Lin, R., Qian, J., Dou, Z. 2021. Design of a flexure-based parallel XY micropositioning stage with millimeter workspace and high bandwidth. *Sensors and Actuators A: Physical* 331, p. 112899.
[41] Wlodarczyk, K., Carter, R., Jahanbakhsh, A., Lopes, A., Mackenzie, M., Maier, R., Hand, D., Maroto-Valer, M. 2018. Rapid laser manufacturing of microfluidic devices from glass substrates. *Micromachines* 9(8), p. 409.
[42] Xu, H., Liu, C., Silberschmidt, V.V., Pramana, S.S., White, T.J., Chen, Z., Sivakumar, M., Acoff, V.L. 2010. A micromechanism study of thermosonic gold wire bonding on aluminum pad. *Journal of Applied Physics* 108(11), p. 113517.
[43] Xu, Z., Song, L., Zhao, Y., Liu, S. 2021. The formation mechanism and effect of amorphous SiO2 on the corrosion behaviour of Fe-Cr-Si ODS alloy in LBE at 550°C. *Corrosion Science* 190, p. 109634.
[44] Yu, C.-M., Lai, K.-K., Chen, K.-S., Chang, T.-C. 2020. Process-quality evaluation for wire bonding with multiple gold wires. *IEEE Access* 8, pp. 106075–106082.
[45] Zhao, C.Q., Xu, X.C., Lu, Y.J., Wu, S.Q., Xu, Z.Y., Huang, T.T., Lin, J.X. 2020. Doping lithium element to enhance compressive strength of β-TCP scaffolds manufactured by 3D printing for bone tissue engineering. *Journal of Alloys and Compounds* 814, p. 152327.
[46] Zhou, H., Chang, A., Fan, J., Cao, J., An, B., Xia, J., Yao, J., Cui, X., Zhang, Y. 2023. Copper wire bonding: A review. *Micromachines* 14(8), p. 1612.
[47] Zhou, H., Chang, A., Fan, J., Cao, J., Zhang, Y., An, B., Xia, J. 2023. Effects of process parameters on bond properties of Ag-2.35Au-0.7Pd-0.2Pt-0.1Cu alloy wire. *Micromachines* 14(8), p. 1587.
[48] Zhu, W., Numkam Fokoua, E.R., Chen, Y., Bradley, T., Petrovich, M.N., Poletti, F., Zhao, M., Richardson, D.J., Slavík, R. 2019. Temperature insensitive fiber interferometry. *Optics Letters* 44(11), p. 2768.

Index

S

T

For Product Safety Concerns and Information please contact our EU representative GPSR@taylorandfrancis.com
Taylor & Francis Verlag GmbH, Kaufingerstraße 24, 80331 München, Germany

www.ingramcontent.com/pod-product-compliance
Ingram Content Group UK Ltd.
Pitfield, Milton Keynes, MK11 3LW, UK
UKHW020204200926
475702UK00004B/607
9781032796925